Quality Control in Analytical Chemistry

CHEMICAL ANALYSIS

A SERIES OF MONOGRAPHS ON
ANALYTICAL CHEMISTRY AND ITS APPLICATIONS

Editor
J. D. WINEFORDNER
Editor Emeritus: **I. M. KOLTHOFF**

VOLUME 60

A WILEY-INTERSCIENCE PUBLICATION

JOHN WILEY & SONS, INC.

New York / Chichester / Brisbane / Toronto / Singapore

Quality Control in Analytical Chemistry

Second Edition

G. KATEMAN
L. BUYDENS

Department of Analytical Chemistry
Catholic University of Nijmegen
The Netherlands

A WILEY-INTERSCIENCE PUBLICATION

JOHN WILEY & SONS, INC.

New York / Chichester / Brisbane / Toronto / Singapore

Library of Congress Cataloging in Publication Data:

Kateman, G.
    Quality control in analytical chemistry / by G. Kateman and L.
Buydens.—2nd ed.
        p.   cm.—(Chemical analysis ;   v. 60)
    "A Wiley-Interscience publication."
    Includes index.
    ISBN 0-471-55777-3
    1. Chemistry, Analytic—Quality control.   I. Buydens, L., 1955–
II. Title. III. Series.
QD75.4.Q34K38   1993
543—dc20   93-12519

CONTENTS

PREFACE

The view that analytical chemistry is interwoven with the objective and the object of analytical tests and that this interaction influences the way analytical quality should be improved was quite new twelve years ago when the first edition of *Quality Control in Analytical Chemistry* was written.

That "Chemometrics can be an important tool for Quality Control in Analytical Chemistry comprising the whole analytical process, from sampling up to organization," was not widely accepted. The change in thinking that has occurred since then and the rapid development of chemometrics have made a revision of the original text desirable. Building on our experience in teaching chemometrics and qualimetrics we have added a number of modern techniques, discarded some obsolete ones, and introduced the latest developments in the time-proven techniques that were already available.

We think that chemometrics deals with all aspects of analytical chemistry in its widest sense and that chemometrics does not comprise only mathematical techniques but all techniques that can help to improve the analytical process. As this view is not fully accepted by chemometricians we refrain from using the term "chemometrics" as much as possible. We think that "qualimetrics" can be a description of the field using chemometrics (in our definition) for quality control in analytical chemistry. An alternative title for the book could be "Qualimetrics."

The basic concept has not changed. New developments have been treated, especially the influence of new mathematical techniques on data treatment (e.g., partial least squares), the possibilities of artificial intelligence in data interpretation and quality control, and the use of simulation experiments and queuing theory for laboratory organization and quality improvement. The list of references at the end of the volume gives all relevant modern literature in books and outstanding journals.

G. KATEMAN

Nijmegen
May 1993

L. BUYDENS

ix

PREFACE TO THE FIRST EDITION

As a manager of a control laboratory of a large chemical plant, one of the authors of this book was continuously confronted with questions on quality, the quality of our product as compared with that of our competitors and the quality of the products of our suppliers. However, the questions to answer were always the following:

1. What is the quality of our (analytical) work?
2. Do we provide the people in the factory and the marketing people with the right answers to their questions?
3. Do they comprehend, trust, and use our data?

In our opinion every analytical chemist should be asked these questions, or he should ask them himself, and be prepared to answer them.

To answer these questions and to improve the quality of analytical chemistry requires a wide view of how quality can be affected. This is the reason that we elaborate on so many different techniques in our curriculum for analytical chemists at the Catholic University of Nijmegen, that we do research in some unusual techniques such as "queueing," and that we wrote this book.

Quality can be defined, and is defined in this book, as "the (numerical) value of a set of desired properties of an object (i.e., chemical analysis)." Thus quality is not an absolute property but depends on the definition of "desired."

In analytical chemistry quality can be associated with accuracy and/or reproducibility. Another quality criterion can be speed, cost, or information flow per dollar. Chemical analysis is always associated with an object to be analyzed. Furthermore, the analysis always has a purpose: to control, to describe, or to monitor the object. So a way of quantifying quality can be to answer the question; how well is the subject of analysis described in view of the applicability of the results?

In this book much attention is paid to the description of the work of the analyst—not a description of the various analytical chemical hardware and methods, but an explanation of how the result of the analysis is affected by the various stages of the analytical procedure:

sampling, analysis, data handling, and, when aiming for a specific goal, control, description, or monitoring. When the reader has learned how quality in chemical analysis may be affected in a great variety of practical situations, it will be quite easy for him to control the quality by optimizing a limited number of selected parameters.

Of course the quality level of analytical chemical work is determined by factors such as the expertise of the analyst, the quality of the description of the analytical procedure, and the performance of the analytical instruments. However, to attain a high standard, to convince oneself and others that the work cannot be done better, and to make clear in quantitative terms how good the quality of the work is, one needs methods to measure and control the quality of analytical chemical work.

G. KATEMAN
F. W. PIJPERS

Nijmegen
January 1981

CHEMICAL ANALYSIS

A SERIES OF MONOGRAPHS ON
ANALYTICAL CHEMISTRY AND ITS APPLICATIONS

J. D. Winefordner, *Series Editor*
I. M. Kolthoff, *Editor Emeritus*

CHAPTER

1

INTRODUCTION

Quality control in chemical analysis is an old concept. As long as there has been chemical analysis, there has been the need to control the quality of its performance. The management of a laboratory has a need to estimate the quality of its efforts and, if quality does not meet the standards, to improve it. This book describes techniques that can be applied to improve the quality of analytical methods, sampling, and data handling. Often the results of these techniques can be used as a criterion for quality.

This leads to several questions. Why do laboratories want to improve their quality level, and what happens if they don't? What quality parameters can be improved, and which one should be improved first? How can quality be improved or controlled, and why is it so difficult to keep quality on a steady, high level? Quality can nearly always be improved, but at the expense of much ingenuity, work, and money. Sometimes the efforts are required to reach a goal; sometimes reasons such as survival of the fittest are at stake. It would be desirable if some criterion could be applied that allows the optimization of quality; that is, the sum of the yield of the object of analysis and the negative yield (cost) of the analysis itself should be maximal. This yield may be expressed in terms of money, but, though money is a good and dependable yardstick, it is not the only one. Efforts in human ingenuity, although these are difficult to measure, could be the parameter of optimization. The amount of manpower or the level of sophistication of the instruments required can serve as a measure of ingenuity.

At this time there is no universal standard in operation. For a long time, accuracy and precision were the only quality parameters in use. These could not be optimized since it was impossible to estimate the yield and the cost corresponding to the various levels of accuracy, for example.

The gradual change in attitude toward the concept of quality control in chemical analysis has only had a short history, though the first applications of quality control date back to the beginning of the century. In 1908, Gosset (GO 08), an amateur of statistics and a professional chemist, published a paper on the error of a mean under his pen name, "Student." He introduced a parameter known since then as Student's-t. It allows analytical chemists (and others) to compare quantitatively the results of two items, such as methods of analysis or samples. Here the quality parameter in use is the level of probability that the two means match. Another statistical parameter that could be used is the standard deviation. This quality parameter combined with Fisher's F-test indicates the probability that the scatter in the results of a particular experiment is better or worse than the scatter found in a very reproducible standard series of analytical results. Although these quality parameters were known, they were not in common use among analytical chemists. In 1928 Baule and Benedetti-Pichler (BA 28) published a paper on the minimum size of samples of inhomogeneous lots in which they indicated the quality parameters for sampling.

The first major impact on quality control in analytical chemistry was caused by the introduction of the control chart, familiarized by Shewhart in 1931 (SH 31). Such a chart allows the continuous supervision of the quality parameters accuracy and precision in an analytical laboratory. It enables corrective measures when a control line is crossed. A serious drawback of this method is the position on the control lines that is arbitrarily chosen.

After 1940 a host of publications appeared on the application of statistical techniques in analytical chemistry. However, most of these were directed to the application of statistics in the validation of analytical results. The main purpose of such statistical validation is the determination of the probability that certain effects in objects or processes can be detected by analytical chemical results. In analytical chemistry the application of statistical techniques for the control of analytical procedures was restricted to the comparison of analytical procedures, without establishing more quality parameters for analytical methods than standard deviation and variance. In 1963 van der Grinten (GR 63) published some simplified equations that were derived from the rules of Norbert Wiener on control theory. This allowed the quantitative evaluation of the quality of measurements with respect to the object that was measured. In the years that followed van der Grinten together with Kateman explored the possibilities of the quality parameter *measurability* at the chemical works of DSM in the Netherlands.

To each component in a chemical process that was sampled and analyzed one could attribute a quality number when some parameters of the process (time constant and standard deviation of the uncontrolled process) were known. In 1971 Leemans (LE 71) published some of the results in the analytical literature. It became clear that, apart from accuracy and reproducibility, the speed of analysis, the frequency of sampling, and their merit for the application of the results could also be quantified and optimized. In 1972 the Chemometrics Society was founded with the aim of tying together the diverse research in statistical, mathematical, and artificial intelligence techniques in analytical chemistry. Many of the developed techniques can be used in quality control. These developments were not the only ones that caused a new view of quality control in chemical analysis. In 1972–1974 some incidents occurred involving certain research laboratories in the United States. The validity of a number of their study reports was questionable. Because of this, the FDA (U.S. Food and Drug Administration) in 1976 issued "Good Laboratory Practice" (GLP) regulations. In 1979 the final regulations came into force. GLP concerns the organization and the circumstances under which laboratory research is planned, conducted, monitored, registered, and reported. Since then, a number of other governmental authorities in the United States and other countries have issued similar regulations. The OECD (Organization for Economic Cooperation and Development) Principles of Good Laboratory Practice resulted from these activities in a European context.

A consequence of the developments mentioned above is the growing interest in the development and application of quality control and quality assurance programs also in those laboratories that do not fall under GLP regulations. The objective is to organize the laboratory in such a way that the number of errors and mistakes is reduced to a minimum. This means that a thorough quality system has to be established. A substantial number of quality management and quality assurance standards have been issued all over the world. A great deal of effort has been put into the harmonization to the different standards. The basis for this is the ISO-9000 series of standards (IN 87, OE 82).

Around the year 1970 a discussion of the necessity of analytical chemistry arose—mainly in the American analytical literature—that culminated in the rhetorical question, "Analytical chemistry, a fading discipline?" (FI 70). The discussion was pursued in Europe and resulted in a series of definitions. These were formulated by Kateman and Dijkstra (KA 79) within three categories of definitions:

1. Analytical chemistry[1] produces information by application of available analytical procedures in order to characterize matter by its chemical composition.
2. Analytical chemistry[1] studies the process of gathering information by using principles of several disciplines in order to characterize matter or systems.
3. Analytical chemistry[1] produces strategies for obtaining information by the optimal use of available procedures in order to characterize matter or systems.

In a way these categories correspond to three distinct nonhierarchical levels in analytical chemistry. The first level refers to the actual production of analytical results. For this production instruments, procedures, and skilled personnel are required. The second level covers the research and development of analytical procedures. The third can be considered as an organizational level. It comprises the interaction between humans and machines, including communication as well as the optimal use of the tools available for producing information. In practice, of course, these levels are interwoven, but the division seemed to be indispensable for a new view on quality control.

In Europe the Arbeitskreis "Automation in der Analyse" (AR 73) formulated the view that analytical chemistry should be considered as a "black box," a system with known input and output and rules interconnecting them. According to this point of view, operations research, the science that studies this kind of system, could be applied to analytical chemistry. Massart (MA 75) presented the first survey of research using this new approach. According to this approach, the application of operations research techniques to analytical chemistry indicates some new quality parameters:

- The degree of optimization of a procedure, if quantitative optimization methods are used
- Information as a measure of quality of a qualitative method

---

[1] Or analytical chemists.

- Waiting time as a measure of planning and routing in analytical chemical laboratories

A third phenomenon, parallel to the use of control theoretical concepts and operations research, was the introduction in analytical chemistry of integrated systems of statistics and correlation analysis, comprising techniques such as multivariate analysis and pattern recognition. These systems greatly improved the possibilities of using multivariate data but introduced new sources of error. Last but not least should be mentioned techniques from artificial intelligence and in particular expert systems. These allow the incorporation of expertise, statistics, method selection, etc., into analytical instruments or analytical procedures and in this way improving quality.

Although the methods mentioned here sometimes are grouped under the heading "chemometrics" (KO 75), it does not seem justifiable to use the term in this book since we confine ourselves to the problem of quality control. This means that many of the techniques covered by chemometrics can be used, but only with one purpose in mind: to help in the validation of analytical procedures and organizations, that is, to establish quality in the broadest sense of the word. The following references are books on chemometrics that contain valuable information: BR 90, DA 84, KA 81, KO 77, KO 84, KU 85, MA 78a, MA 78b, MA 88b, MI 88, SH 86; research papers in that field are published in many journals, but mainly in *Chemometrics and Intelligent Laboratory Systems, Journal of Chemometrics, Trends in Analytical Chemistry, Analytica Chimica Acta, Analytical Chemistry, Fresenius' Journal of Analytical Chemistry, The Analyst* and *Journal of Chemical Information and Computer Systems. Analytical Chemistry* publishes biennial reviews on chemometrics.

A totally new approach is the "total concept." In this approach all aspects of an analytical procedure are taken care of by a sophisticated expert system, looking for the best method of analysis on the basis of a first guess, providing descriptions of sampling methods and the like, performing an optimal calibration, and in the end, after a robot-driven analysis, writing a complete protocol that describes all parameters of the analysis and the reasons for their application. The goal seems far fetched, but many groups are working in this direction (GO 88, KI 89).

The development in clinical chemistry was parallel and apparently independent. The number of analyses in this field doubled every 5 years;

thus clinical chemists were confronted with the need to cope with this problem. Here the first line of attack was automation and, as is clearly visible when one visits a clinical laboratory, progress in this field was great. Clinical chemists became aware of the quality problem quite early, even before automation was begun, when confronted with the enormous differences in clinical chemical data between various laboratories. Clinical chemists came to recognize that their laboratories are in many ways similar to a manufacturing plant: reagents and samples are received and subjected to a variety of manipulations using specialized tools and instruments. The final product is the analytical data that are produced.

In 1947 Belk and Sunderman (BE 47) introduced the Shewhart chart as a means to control the product. The conclusions of Belk and Sunderman based on their pioneer survey were later confirmed by the results of similar studies. All revealed a wide range of reported values for the analysis of identical specimens among a group of laboratories. In 1950 a formal intralaboratory quality control system based on estimation of accuracy and precision was first described by Levey and Jennings (LE 50). Along this line many nationwide, interlaboratory control systems have been established, some under governmental supervision, others under the supervision of boards founded by clinical chemists. However, at present there is no system available that measures quality unambiguously and creates a uniform quality of clinical chemical measurements. The vast problems caused by the increasing work load also gave rise to research on information content, cost, and waiting times (VA 74). This research seems to be independent of the analytical operations research trend described earlier.

One way of defining quality is by stating it to be "the [numerical] value of a set of desired properties" or "the similarity between wanted and estimated properties." In order to control the quality of analytical chemical methods and results, one must define the set of desired properties of the analysis as well as the way to quantify these and the standards to be set.

For practical reasons it is sufficient to consider only those properties that can be quantified and for which standards can be set. Of course it might be possible that more properties are desired, but if they cannot be quantified it is useless to mention them in the present context.

The process of measuring and varying analytical quality can be compared with the control of a technological process. In principle the same procedures can be applied and many of the statistical and control theoretical procedures described elsewhere could be of use. However, an analytical laboratory is far more difficult to control than a chemical process. Some reasons for this can be mentioned:

- Only a few quality parameters are known and fully understood. Measurement of these parameters is often time consuming and expensive. The number of measurements that can be used for control is often very low, so that "normal" statistics do not apply.
- Most analytical chemical laboratories are relatively labor intensive, and human effort effects the quality of the work considerably. Quality defects caused by human error are very difficult to control and cannot be coped with completely by control theory.
- The results of analytical chemical measurements affect more than just the institute or factory for which the laboratory is working. Products are bought and sold all over the world, and water and air pollution is transported across boundaries of countries and continents, for example. Therefore the results of analytical laboratories should meet not only laboratory but also national and international standards.
- In analytical laboratories a comparatively large part of the staff consists of highly and very highly skilled and competent workers.

These factors, in combination with the ease and low cost of changing an analytical procedure, are responsible for the astonishing small number of exactly corresponding analytical procedures. Both comparison and control of incomparable procedures on a large scale are difficult.

Nevertheless the analytical world has adopted some control instruments that can cope with the situation and use the unique structure of that world. This control mechanism is based mainly on three assumptions:

1. An analyst is able to perform his/her own control measurements and evaluate them.
2. A laboratory worker can and is willing to improve the quality of his/her work.
3. A system should be established that gives the analyst the right and the duty to perform tasks 1 and 2.

This combination, known as Good Laboratory Practice (GLP), seems to be sufficient for control in practical situations.

For a more detailed consideration it suffices to distinguish three levels in analytical work that should be controlled. The first level pertains to the unit operation. Some of the operations, for example, sampling, measuring, and data processing, can be evaluated by quantifying desired properties, say, accuracy or reproducibility. The second level pertains to the procedure itself. Many of the quality parameters described in this book are on the procedure level, for example, sampling frequency, limit of detection, and sensitivity. The third level belongs to the laboratory considered as a unit. When the performance of analytical procedures is evaluated, not only the results of various methods, workers, and instruments are compared but also the results of other laboratories where comparable procedures are applied. Examples of quality parameters at this level are accuracy, cost, speed, and time lag between question and answer.

The next three chapters of this book cover the three parts of the analytical process: sampling, analysis, and data handling. As will be seen, the methods used to describe quality are not often applied exclusively to one of these areas. Frequently the techniques have a much wider scope and therefore the concept of measurability, for example, will be found in the section of sampling as well as in the section on analysis. We feel that the control mechanism that works in many laboratories is reliance on the craftsmanship and responsibility of the analyst. However, certain incidents in the past have proved that laboratory results can not always be trusted. As indicated earlier, GLP tries to prevent errors by prescribing a strict organizational scheme and promoting quality awareness. In this book we do not rely on strict rules but try to accommodate the analyst with a set of tools, to be applied when and where they are wanted. Since these tools are intended for analytical chemists operating in common practical situations of analytical laboratories, in charge of performing analyses for control and research, we refrain from giving rigorous derivations of the equations presented. Interested readers will find the theoretical background in the original papers cited.

The concept of quality in chemical analysis has not been treated in detail in the literature up to now. Therefore it is not possible to give here a complete treatment of all parameters that influence quality. However, it seems possible to improve quality where the influence of the parameters is known. Another problem is the quality level that is required. Sometimes the level is set by the use that has to be made of the analytical results, and sometimes an optimum exists, depending on other variables. In Section 1.1 a survey is given of quality criteria in chemical analysis.

1.1. QUALITY IN CHEMICAL ANALYSIS

In this section we try to define various quality parameters that are discussed in this book. Because quality in general can be defined as the value of a set of desired properties, it is necessary to describe the parameter that denotes the quality of a certain property of chemical analysis. Furthermore, how quality can be influenced and what features determine the ultimate or optimal quality are discussed.

1.1.1.   Criteria

The quality of a *sample* is given by the similarity between sample and object. Its magnitude can be expressed by the standard deviation in the composition of many samples, or otherwise as the bias in the composition of a sample compared with the "true" value of the composition of the object. Sample quality is influenced by the composition and inhomogeneity of the object, the sample size, and the method of sampling (Sections 2.4, 2.5, and 2.6). The upper limit of the quality is set by the precision of the method of analysis that is used to analyze the sample. In addition, the quality of a sample can be influenced adversely by inadequate subsampling, decomposition, and contamination (Section 2.7).

The quality of the *sampling method* is given by the similarity between the reconstruction of the composition of an object and the object itself, as far as it is influenced by the sampling strategy. It depends on the characteristics of the object and the purpose of the reconstruction (Sections 2.3 and 2.4).

For a mere description of the object, the quality can be expressed in the sample quality (Sections 2.6.1 and 2.6.2).

For monitoring of the object, the sampling quality can be expressed as the probability that a threshold crossing will be detected (Section 2.6.3).

For object control the quality can be expressed as controllability or measurability (Section 2.6.4).

Another method for measuring sampling quality is to estimate the effort (expressed in the number of samples to be taken) required to obtain a specified reconstruction error. Here sampling quality is influenced by the inhomogeneity, internal correlation, and size of the object, and the number, spacing, and size of the samples. The maximum quality that may be achieved is set by a combination of these factors. The optimal quality is set by the yield of the reconstruction and the cost of sampling. Moreover, the quality of sampling can be influenced adversely by inadequate sample handling, analysis, and data processing (Sections 2.7 and 3.2.5 and Chapter 4).

The *limit of detection* is a quality parameter pertaining to analysis. It gives the minimum concentration of a component that can be detected. It is influenced by the absolute value of the blank, the standard deviation of the method of analysis, and a safety factor (Section 3.2.1). The lowest possible limit of detection is set by the characteristics of the method.

The *sensitivity* is another quality parameter of the analysis. It gives the change of a signal upon a change of concentration (Section 3.2.2). The sensitivity of a method is a practical quality measure, since it pertains to the ease of detection. However, both the detection of a difference in concentration between two samples and the limit of detection are ultimately governed by the precision of the analytical method and not by the sensitivity. The optimal value of the sensitivity should be such that the standard deviation of the method can be measured.

Selectivity and *specificity* are quality parameters of a method of analysis. They can be expressed as the ratio of sensitivities of the method for various components to be measured in the sample. The optimal value of these characteristics is set by the number of measurements required to estimate the composition of the sample (Sections 3.2.8 and 3.2.9).

Safety is an important but ill-defined quality parameter of a method of analysis. Here hardly anything is known about quality criteria. Perhaps the cost of measures to be taken to satisfy safety regulations can be used to measure safety as a quality parameter (Section 3.2.3). The optimum is determined by the balancing of adverse consequences and precaution costs.

Cost is an important quality parameter of an analytical method. However, there are hardly any publications on the subject. Cost can be expressed in money or in other measures, such as manpower. It is influenced by analysis time, standard of labor, depreciation, and cost of instruments, energy, and reagents (Section 3.2.4). The optimal cost of an analytical method results from a comparison of various methods of analysis with respect to the yield of the object and cost of analysis (Section 2.6.4).

Precision is a quality parameter that, important as it may be, is not well defined. *Analytical Chemistry* (AN 75) proposes the following: "Precision refers to the reproducibility of measurement within a set, that is, to the scatter or dispersion of a set about its central value." The term *set* itself is defined as referring to a number of independent replicate measurements of the same property. The International Organization for Standardization (ISO) (IN 66) applies two descriptions of precision: (1) the *reproducibility*, the closeness of agreement between individual results obtained with the same method but under different conditions, and (2) the *repeatability*, the closeness of agreement between successive results obtained with the same method and under the same conditions.

Both descriptions of precision use the same measure, the standard deviation of the measurements. Apart from the influence of external conditions, the precision is influenced by the number of replicate measurements (Section 3.2.6). The ultimate precision of analysis is determined by the precision of the method and the quality of the sample and sampling (Sections 2.6.2, 2.6.3 and 2.6.4). The optimal precision is determined by cost, sample, sampling, and aim of the analysis (Sections 2.6.4 and 3.2.4).

Accuracy is an even less well defined quality criterion. *Analytical Chemistry* (AN 75) states: "Accuracy normally refers to the difference (error or bias) between the mean of the set of results and the value which is accepted as the true or correct value for the quantity measured." Because most analytical methods claim to give the true value, accuracy as such is seldom used as a quality criterion (Section 3.2.7). Often the difference between the mean of the set of results and the mean of a much larger set is taken for accuracy (Section 3.4.3 and 4.6).

Traceability is a vague but important aspect of analytical quality. Traceability means that all aspects of quality are defined or can be coupled to some recognizable standard. Measurements should be based on sound chemical or physical principles, described in the open literature and thoroughly tested. Standard materials or procedures should be used for calibration, and the specifications of these materials should be available. The same holds for calibration procedures. An important part is the traceability of software that is incorporated in analytical instruments or commercial software that has no description of the underlying algorithms. This part of quality control is still in an embryonic stage.

Information can be considered as a quality parameter for quantitative and qualitative analysis and for data retrieval. Information is defined as the difference in uncertainty before and after an experiment. It is expressed as the binary logarithm of this uncertainty and measured in bits. The information content of a qualitative analytical method is governed by the selectivity and specificity of the method and the a priori knowledge of the occurrence of the sought component. The information content of a quantitative analytical method is a function of the standard deviation of the method and the a priori knowledge of the range of compositions that can be expected. The information content of a retrieval method is governed by the selectivity of the method and the chance of occurrence of the sought item. Therefore information cannot be a quality criterion for a method as such. It can be applied only in a specified situation. Correlation of items makes the a priori information partly predictable and thus diminishes the information content of the experiment (Section 4.2).

1.1.2. Methods

Curve fitting is not a quality parameter, but it may enhance the quality of measured data considerably for two reasons:

1. Interpolating the gap between data allows easier interpretation of many phenomena, although the information does not increase.
2. A most important use of curve fitting is in the unraveling of mixed phenomena, for example, spectra (Section 1.4).

Multivariate analysis may be a tool in establishing new quality criteria. The first application may be in discovering patterns in the data produced by analytical chemists. This enhances the value of these results. A measure for this effect is not known, but a "patterned" result seems to increase the quality of the analytical results.

Optimization is not a quality parameter. However, it may be applied in optimizing analytical methods and thus improving quality. Perhaps the difference between a method as such and a method optimized with respect to a number of quality parameters could be a quality criterion (Section 4.9).

1.2. QUALITY CONTROL

Quality of chemical analysis can be improved or kept on a steady level by controlling the parameters that influence quality. The most direct way to do this is to control the parameters that are given by the equations that describe quality quantitatively. Although many quality parameters cannot be described quantitatively, quality may be influenced. A number of techniques that influence or may influence quality are described in this book.

Sampling as a means to influence quality was described in Section 1.1. *Handling of samples* can be of great importance for some quality aspects of a sample. Homogenizing influences the quality of the subsample (Section 2.7.1), and so does sample reduction (Section 2.7.2). Preservation and labeling influence both the identity and integrity of the sample, an immeasurable quality aspect (Section 2.7.3).

The *selection* of an analytical method influences many quality aspects. The selection may be based on one or more quality criteria, and therefore these aspects should be weighed a priori (Section 3.3). Techniques for selection are, for example, the ruggedness test (Section 3.4.1), the ranking test (Section 3.4.4), pattern recognition (Section 4.8), optimization (Section 4.9), and artificial intelligence (Section 4.8.7).

Repeatability is the measure of variations between test results of successive tests of the same sample carried out under the same conditions, i.e., the same test method, the same operator, the same testing equipment, the same laboratory, and a short interval of time.

Reproducibility is the term used to express the measure of variations between test results obtained with the same test method on identical test samples under different conditions, i.e., different operators, different testing equipment, different laboratories and/or different time.

Precision control and *accuracy control* can be performed in several ways. Control charts are means to trace quality in time and indicate when intervention is required (Section 3.4.2). Round-robin tests may

reveal shortcomings in precision and accuracy and, though complicated psychological and organizational influences play a role, quality
can improve after a test (Section 3.4.3). Sequential analysis (Section 3.5)
as well as analysis of variance (Section 4.6) may be used to control
precision and accuracy. Neither method influences the quality parameters, but both may be of help in deciding which method is preferred.
A number of methods can be used to improve precision by removing or
diminishing irrelevant data (noise). Some techniques treated are curve
fitting and smoothing (Section 4.4), Kalman filtering, and deconvolut-
ing filtering (Section 4.3). To improve data handling and quality control by statistical means, often a mathematical transformation of data
may be helpful (Section 4.1.2).

Planning may improve some quality aspects, for example, speed and
cost; by planning over a longer time span, other quality aspects may be
influenced by research. Speed may be improved by the optimal use of
available personnel and instruments. Dynamic modeling, expert systems, and queuing theory provide tools for simulation on laboratory
models (Section 5.1.1). Gantt charts and network diagrams provide the
analyst with simple means to avoid dead times (Section 5.1.3).

The *organization* of a laboratory influences the quality of chemical
analysis in many ways, mostly complicated and obscure. The most
evident influence is found on speed and cost, but accuracy and precision and the entire subject of quality control may be advantageously or
adversely influenced by the organization of analytical work. Therefore,
a section on organizational aspects of quality control is included
(Section 5.2).

SAMPLING

2.1. INTRODUCTION

Sampling is often called the basis of analysis: "The analytical result is never better than the sample it is based on." In our opinion, every part of the analytical procedure is important, because the final result is influenced by mistakes or added noise in every part of the procedure. Of course, this statement simply emphasizes the need to control the sampling as well as any other part of the analytical work.

The purpose of sampling is to provide for a specific aim of the client a part of the object that is representative of it and suitable for analysis. The quality of the sampling depends on many parameters, governed by the properties of the object and of the analytical procedure. The properties of the object can be described in terms of physical structure, size, and inhomogeneity in space and time.

As a rule the analytical chemist cannot or will not use the whole object in his/her analysis machine but uses only a small part of the object. In practice this fraction can be very small: the amount of material introduced in the analytical method rarely exceeds 1 g but as a rule is not more than 0.01–0.1 g. This can be part of a shipload of ore, say, 10^{11} g, or a river transporting 10^{13}–10^{15} g water/day. In many instances this fraction is larger, but a fraction of the object to be analyzed of 10^{-4}–10^{-5} is common practice.

The implications of such a small fraction of the object to be investigated are enormous. The *sample*, as we will call this fraction, must fulfill a series of expectations before it may be called a sample and used as such. It must—

- Represent faithfully the properties under investigation: composition, color, crystal type, etc.

- Be of a size that can be handled by the sampler
- Be of a size than can be handled by the analyst, say, from 0.001 to 1.0 g
- Keep the properties the object had at the time of sampling, or change its properties in the same way as the object
- Be suitable to give the information that the client wants, e.g., mean composition, or composition as a function of time or place
- Keep its identity throughout the whole procedure of transport and analysis

To satisfy these demands, the analyst can use the results of much theoretical and practical work from other disciplines.

Statistics and probability theory have provided the analyst with the theoretical framework that predicts the uncertainties in estimating properties of populations when only a part of the population is available for investigation. Unfortunately this theory is not well suited to analytical sampling. Mathematical samples have no mass, do not segregate or detoriate, are cheap, and are derived from populations with nicely modeled composition, e.g., a Gaussian distribution of independent items. In practice the analyst does not know the type of distribution of the composition and usually has to do with correlations within the object, and the sample or the number of samples must be small, as a sample or sampling is expensive.

The properties of the analytical procedure can be described in terms of minimum and maximum sample size, physical structure of the sample, homogeneity of the sample in space, and stability in time, for example.

The requirements set by the object and by the analyst must be fulfilled, so the sampler has to optimize the sampling parameters such as size, time or distance spacing, number of subsamples, cost, mixing, and dividing.

In this chapter we deal with these parameters, their influence on the quality of the sample, and the procedures to be used.

For a complete review of sampling, including sampling techniques, the reader should consult the literature (e.g., KR 84).

2.2. DEFINITIONS

2.2.1. Quality of Sampling

A definition of *quality* is "the [numerical] value of a set of desired properties," or "the similarity between wanted and estimated properties."

The estimation of the value of the properties is the aim of the analytical procedure. A feature of the object to be known is a list of the desired properties, for these determine the way of sampling.

The object can have many properties, for example, composition, particle size, color, and taste, but only a selection of these properties is required to describe the object quality. In assessing the quality of coal it makes a difference whether the desired property is "ash content" or "particle size distribution," though in both cases the analyst as well as the sampler must consider the proper sampling of the different fractions of different particle size.

One sometimes comes across the term *total analysis*. Most often this refers to a known list of desired properties, but sometimes "total analysis" simply means as many properties as can be estimated.

A stain of paint found on a shirt has quality properties that are different from those of the can of paint from which it comes. Though it is possible to define object quality in this way, the implication is that the analyst or sampler must define the object quality afterward: analyzing the stain with respect to pigment composition, binder, color, or degree of polymerization, for example, is possible by analyzing a proper sample of the stain. The object qualities "weight" or "solvent content" can be established only by sampling the stain in a different way.

2.2.2. The Object

Before sampling and analysis some aims must be set: what is the object to be described and what quality parameters[1] are wanted? We define an *object* here as "the entity to be described."

An object can be a fertilizer granule as well as a bag or truckload of fertilizer, or even the quantity produced last week. An object can be a

---

[1] To avoid confusion, in this chapter the word *quality* is not used on its own, as in normal use, but always coupled to an adjective, as in *sampling quality* or *object quality*.

river at its full length during one year, or the water flowing past a given point during a given period of time, or the water at one point at a given moment.

As a rule the object is sufficiently described by the four coordinates of space and time, though in practice other labels may be more useful. Other variables that are required in order to describe the object are the state variables temperature and pressure, for often the sample cannot be kept in the same condition.

2.2.3. The Sample

Before describing "sample qualities," we define sampling and samples. A useful working definition of a *sample* is "a representative part of the object to be analyzed." Because representativeness is needed only for the object quality parameters, it can also be stated, "A sample is part of the object selected in such a way that it possesses the desired properties of the object." As shown later the sample can resemble the object only to a certain degree, but because this degree is set by the method of analysis, it can be omitted in the definition of sample.

A sample the size of a truckload can be a good sample—that is, representative of the object—but it cannot be used in the laboratory. A very small sample can be as good a sample as a larger one, but perhaps it is impossible to handle. To make the definition useful in practice the following criterion must be added: "The sample must have such dimensions that it can be analyzed."

Another criterion that has to be added is, "The sample must keep the properties the object had at the time of sampling, or change its properties in the same way as the object." Thus deterioration of the sample through exposure to the air, the sample container, or microorganisms, for example, should be avoided.

The sample procedure, or *sampling*, is a succession of steps performed on the object ensuring that a sample possesses the specified sample quality.

2.3. PURPOSE OF SAMPLING

Because the particular sample procedure depends on the purpose of sampling, the various purposes must be delineated.

2.3.1. Description

One purpose of sampling may be the collection of a part of the object that is sufficient for the description of the gross composition of the object, for example, sampling lots of a manufactured product, lots of raw material, or the mean state of a process. Here it is desirable to collect a sample that has the minimum size set by the condition of representativeness or demanded by handling.

Another purpose of sampling can be the description of the object in detail, for example, the composition of a crystallized rock as a function of space or the composition of the various particles in a mixed pigment. Here it is necessary to know the size and the number of samples or the distance between the samples (sampling frequency).

2.3.2. Threshold Control

Here the purpose of sampling is to collect a quantity of sample sufficient for guarding the object. This situation differs from that of control. In guarding it is often sufficient to know that a threshold value has been reached, or will be reached with a certain probability. If action is required, it can be approximate. When the percentage of active component of a catalyst is monitored, for example, the action is the replacement of the catalyst when the threshold level has been reached or even before. The guarding of the effluent of a factory may have the same characteristics: when a certain level is reached or will be reached within a given time, the effluent stream is stopped.

Another characteristic of threshold control is that in most cases only one level is monitored, sometimes the level pertaining to the high limit threshold and sometimes the low one. This is contrary to the situation met with control, when deviation to either side should be eliminated. In threshold control the quality parameter can be the optimal number of samples required for guarding the process or the minimum number of samples required to predict when the process will reach the threshold value. The time span between the warning that the process will reach the threshold level and the actual crossing of that level also can be a quality parameter.

2.3.3. Control

In industrial practice most of the analytical effort is dedicated to process control. The purpose of such control is to keep a process quality constant (e.g., a manufacturing process), as close to a desired value as is technically possible and economically desirable. Part of the deviation from the set point is caused by random fluctuations of process conditions. These fluctuations cause the composition of the sampled product to vary at random between certain limits. In order to control the fluctuating process, samples must be taken with such frequency and analyzed with such reproducibility and speed that control actions can have the optimal result.

A second purpose of sampling can be the detection of nonrandom deviations: drift or cyclic variations. Special techniques are required to discover such anomalies as soon as possible. These also set the conditions for sampling frequency and sample size.

2.4.   TYPES OF OBJECTS AND SAMPLES

Objects can be divided into three groups according to the scheme of Table 2.1. The discrete or continuous changes in object quality can

Table 2.1. Types of Objects

Object

Homogeneous
No change of quality
   throughout the object

Heterogeneous

Discrete change of
quality throughout
the object

Continuous change of
quality throughout
the object

Homogeneous	Discrete Changes	Continuous Changes
Well-mixed liquids	Ore pellets	Fluids or gases with gradients
Well-mixed gases	Tablets	Mixture of reacting components
Pure metals	Crystallized rocks Suspensions	Granulated materials with granules much smaller than sample size

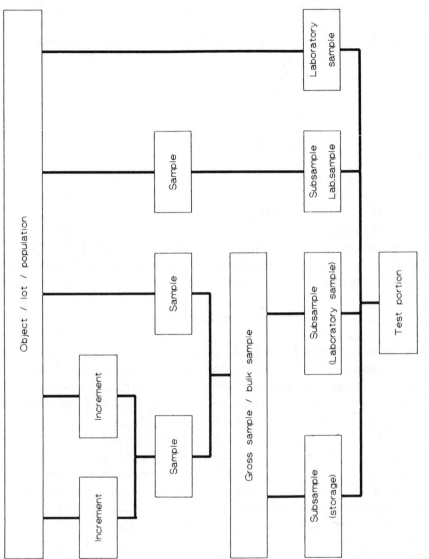

Figure 2.1. Sample nomenclature.

21

manifest themselves in time as well as in space. The composition of the output of a factory changing with time can be seen as a time-related property by sampling the output. It can be seen as a space-related property by sampling the conveyor belt or the warehouse. In many cases such changes are common. Therefore both kinds of changes are treated as equal. Examples of the three types of objects are given in Table 2.1. These types can be distinguished only at the macroscopic level. When objects the size of molecules are considered, all objects are heterogeneous with discrete quality changes (see Figure 2.1).

2.4.1. Homogeneous Objects

The sampling of homogeneous objects poses fewer problems than the sampling of other types. Because the definition of *homogeneous* is "having the same composition at every point and every moment," any increment taken without bias is a sample. It is clear that real homogeneous objects do not pose many problems with regard to sampling. However, true homogeneous objects are rare and homogeneity may be assumed only after verification.

2.4.2. Heterogeneous Objects

Heterogeneous objects can be divided into two subsets: those with discrete changes of the properties, and those with continuous changes. Examples of the first category are ore pellets, tablets, bulk blended fertilizer, and coarse crystallized chemicals. Examples of the second category are larger quantities of fluids and gases, including air, mixtures of reacting compounds, and finely divided granular material.

The second category, samples with continuous changes, can be subdivided into several types. One type of object comprises those with random distribution of the property parameter. This type is rare and exists only when made intentionally. More common is the type that has a correlated distribution of random properties, e.g., the output of a chemical process or the composition of a flowing river. The behavior of the properties can be understood by considering the process as a series of mixing tanks that connect input and output. The degree of correlation can be described by Fourier techniques or better by autocorrelation or semivariance.

A special type of heterogeneous object exhibits cyclic changes of its properties. Frequencies of cyclic variations can be a sign of daily influences as temperature of the environment or shift-to-shift variations. Seasonal frequencies are common in environmental objects like air or surface water.

These several types of heterogeneous objects are seldom found in their pure state; mixed forms with one or two dominating types are most common.

2.4.3. Types of Samples

Often samples are made up from increments or grabs. As long as these increments are real samples, used to build up the sample to increase its size, they are called samples used to build up a gross sample. However, when these increments are parts of the object but not representative of the object, they are not called samples but *sample increments.*

The sample or gross sample can be subdivided into smaller samples. When the properties of these samples are equal to those of the gross sample, or have the required sample quality, they are called subsamples. Note that a subsample by definition is a real sample! A laboratory sample is a subsample or sub-subsample used for analysis. It is the only member of the sample family that must fulfill the added criterion of required dimension.

Other samples, as a rule being subsamples with the same sample quality, can be used for different purposes, such as duplicate analyses, referee analysis, and storage. This leads to the scheme of Figure 2.1 and Table 2.2.

Often a sample is not a separate part of the object but the whole object, for example, very small objects or objects to be analyzed by nondestructive methods. Sometimes a part of the object is declared to be a sample, for example, when analyzing a steel pipe with emission spectroscopy in situ. The same definitions hold, although not all kinds of samples can exist for these cases.

Ku (KU 79) stated that a prerequisite to the development of an efficient analytical strategy is definition of the purpose for which the results are going to be used. This point has been laid down in the recommendation of the ACS Committee on Environmental Improvement (AC 83).

Table 2.2. Glossary of Terms Used in Sampling

Bulk sampling—Sampling of a material that does not consist of discrete, identifiable, constant units, but rather of arbitrary, irregular units.

Gross sample (also called bulk sample, lot sample)—One or more increments of material taken from a larger quantity (lot) or material for assay or record purposes.

Homogeneity—The degree to which a property or substance is randomly distributed throughout a material. Homogeneity depends on the size of the units under consideration. Thus a mixture of two minerals may be inhomogeneous at the molecular or atomic level but homogeneous at the particulate level.

Increment—An individual portion of material collected by a single operation of a sampling device, from parts of a lot separated in time or space. Increments may be either tested individually or combined (composited) and tested as a unit.

Individuals—Conceivable constituent parts of the population.

Laboratory sample—A sample, intended for testing or analysis, prepared from a gross sample or otherwise obtained. The laboratory sample must retain the composition of the gross sample. Often reduction in particle size is necessary in the course of reducing the quantity.

Lot—A quantity of bulk material of similar composition whose properties are under study.

Population—A generic term denoting any finite or infinite collection of individual things, objects, or events in the broadest concept; an aggregate determined by some property that distinguishes things that do and do not belong.

Reduction—The process of preparing one or more subsamples from a sample.

Sample—A portion of a population or lot. It may consist of an individual or groups of individuals.

Segment—A specifically demarked portion of a lot, either actual or hypothetical.

Strata—Segments of a lot that may vary with respect to the property under study.

Subsample—A portion taken from a sample. A laboratory sample may be a subsample of a gross sample; similarly, a test portion may be a subsample of a laboratory sample.

Test portion (also called specimen, test specimen, test unit, aliquot)—That quantity of material of proper size for measurement of the property of interest. Test portions may be taken from the gross sample directly, but often preliminary operations such as mixing or further reduction in particle size are necessary.

Source: Reprinted with permission from B. Kratochvil, D. Wallace, and J. K. Taylor, *Anal. Chem.* **56**(5), 114R (1984). Copyright 1984 American Chemical Society.

An introduction to sampling and influences on sampling can be found in Cochran (CO 77). More general discussions of sampling and sampling statistics are given in the literature (HO 76, KR 81, SM 81), as are several reviews on sampling (IL 85, KR 84, PI 85; a review covering older references is, e.g., KR 82). A software program allowing simulation of many sampling situations and providing calculation algorithms for sampling schemes has been published (KA 85).

2.5. SAMPLE QUALITY

Sample quality is to be distinguished from object quality. Object quality depends on the set of properties that must be estimated in the analytical procedure. Sample quality depends on the set of properties of the sample that leads to a good sample. The primary sample quality parameter is representativeness. Secondary parameters are size, stability, and cost. For changing objects, additional parameters are discriminating power and/or speed.

2.5.1. Representativeness

The sampling procedure must fulfill the condition that analysis (with a required accuracy) of the sample shows no difference in object quality compared to the object itself. This means that two samples of the same object cannot be distinguished. In practice this condition is often not met. Only when the accuracy of the total analysis (including sampling and analysis) is diminished can the aforementioned condition be fulfilled. Two cases can be distinguished: the sampling is prone to random variations, or the sampling causes systematic deviations. In Section 2.6 the connection between reproducibility of the method of analysis and method of sampling is derived.

2.5.2. Systematic Deviations in Sampling

Some causes of systematic deviations in sampling are the following:

1. The number of increments is too small, so that the sample shows a bias. In fact this is a random deviation.

2. The sampling procedure is preferential to one or more object quality parameters. Continuous sampling of air or water can be biased for particles because of the different inertia of the particles compared to the bulk. Sampling of granulated product at a conveyor switch point can miss the fines, which escape as dust. Electrostatically charged particles tend to clog or escape, as can living organisms.

3. The sampling procedure causes alterations in the object. Obvious causes are crushing of granules during sampling, evaporation, and segregation, for example. More complicated are sampling procedures that trigger a reaction in the object, such as catalytic decomposition by the sampling tool, oxidation by oxygen introduced during sampling, fear-induced alterations of zoological samples, and changing reactions of living cells, for instance, the potassium "bleeding" of cells after venopuncture.

4. The sample changes after the sampling procedure and before the analysis. Examples are the oxidation, dehydration, or biological degradation of samples, coagulation, and condensation. A distinct cause is the interaction with the sample container by corrosion, adsorbtion, or catalytic action.

5. The sample is altered intentionally. The results of analysis can have strong economic or psychological repercussions. Because often during the process of sampling humans can or must interfere, and because the sampling is nearer to the involved party than analysis is, sampling is liable to intentional alterations. Bribery, the fear of being accused of malproduction, and the more or less "aesthetic" desire to show the object at its best are some of the more common causes.

2.6. SAMPLING PARAMETERS

Since the sampling procedure affects the sample, the so-called sampling parameters that define this procedure must be considered. The object or process is represented by the distance 0–1 in Figure 2.2. It is an ergodic time series, which means a process that can be represented mathematically by a time series of such a length that the properties of that series (mean, standard deviation, and autocorrelation) are independent of the length. In this time series consecutive numbers can

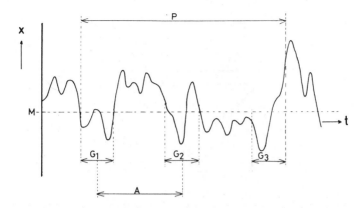

Figure 2.2. Lot size P, grab size G, and distance between samples A as a function of time or distance.

represent, say, composition as a function of time or distance. In practice often a finite part of the process is considered, a nonergodic time series. This finite part, called a lot, can be represented by the distance P, as shown in Figure 2.2.

A sample is the sum of n sample increments, or grabs, taken during a time span G with a frequency $1/A$. In the rest of this chapter the influence of the parameters G, A, and n on the estimation of the composition of the lot P or the process is considered for various types of lots.

2.6.1. Samples for Gross Description

2.6.1.1. Homogeneous Objects

If the object to be sampled is homogeneous, for example, a well-mixed tank or lake, it is obvious that here every sample, whatever its size, is a true copy of the object. Theoretically one sample suffices. Often in practice more samples are needed, for instance, to fulfill the requirement that the sample must have a minimum size. When that sample size is too large, such as in precious *objets d'art*, more samples are needed. The size of the gross sample, set by the requirements of the analysis, can be estimated as follows: Assume the sample weight W and the fraction of the component to be estimated is P ($1\% \rightarrow P = 0.01$; $1\,\text{ppm} \rightarrow P = 10^{-6}$). Now the total amount of the component in the

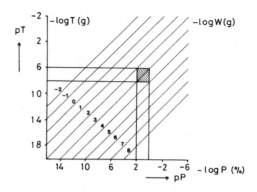

Figure 2.3. Nomogram for the interdependence of sample size W, composition P, and total amount of component T (AR 72).

sample is $T = PW$. A yardstick for sample size estimation is provided by the nomogram of Figure 2.3, which is based on the use of the logarithmic operator $p = -\log$. This implies that $pT = pP + pW$ (AR 72).

2.6.1.2. *Heterogeneous Objects*

The foundations of sampling theory, seen from the point of view of the object, were laid by Baule and Benedetti-Pichler (BA 28), Visman and colleagues (VI 69, VI 71, VI 72), and Gy (GY 82, GY 86).

Their starting point was the object that consists of two or more particulate parts. The particulate composition is responsible for a discontinuous change in composition. A sample must be of sufficient size to represent the mean composition of the object. The cause of difference between an increment of insufficient size and the object is the inevitable statistical error. The composition of the sample, the collection of increments, is given by the mean composition of the object and the standard deviation of this mean. As the standard deviation depends on the number of particles, the size, and the composition of these particles, an equation can be derived that gives the minimum number of particles.

Suppose an object consists of two granular components, A and B; the particles are spherical granules of identical size and density. Now it is possible to predict the composition of an increment. The composition of an increment depends on the size of the increment, for two different reasons. The particulate composition is responsible for a discontinuous change in composition. If the composition is 50% A and 50% B, say, an increment of one granule will have the composition either 100% A or 100% B. An increment of two granules can have the composition 0, 50, or 100% A; a three-particle increment can be 0, 33, 67, or 100% A.

The cause of difference between the increment and the object is the inevitable statistical error in taking an increment from the object. The composition of the grab, the collection of increments that must ultimately constitute a sample, is given by m and s, with m being the mean composition of the object and s the standard deviation of this mean. From statistics it is known that $s = (P_A P_B n)^{1/2}$ particles or

$$s_g = \frac{1}{n}(P_A P_B n)^{1/2} \times 100 = 100 \left(\frac{P_A P_B}{n}\right)^{1/2} \% \qquad (2.1)$$

where s_g = standard deviation of the increment
P_A = fraction of component A in the object
$P_B = 1 - P_A$ = fraction of component B in the object
n = number of particles

Assume the same composition as before, 50% A, 50% B, $P_A = 0.50$, and $P_B = 0.50$. An increment with $n = 10$ will have a composition

$$m = 50\% \text{ A}; \qquad s_g = 100 \frac{(0.5 \times 0.5)^{1/2}}{10} = 16 \qquad (2.2)$$

Two out of three increments will have a composition between $50 \pm 16\%$, or between 3 and 7 particles A. The probability that an increment is found outside this region is $1:3$. In fact, the probability of finding an increment with the composition 10 particles A or B is not negligible: 1% of the increments will have a composition of 0 or 10 particles of one kind.

From eq. 2.1 it follows that the chance of a deviating composition diminishes with increasing n. If the increment is considered to be representative of the object, when there is no means to distinguish this increment from another one taken from the object, a method of analysis is required that has an ultimate accuracy larger than the standard deviation of the increment. When the accuracy of the analysis is

$$s_a = s_A/N^{1/2}$$

where s_a = standard deviation of the analysis
s_A = standard deviation of the method of analysis
N = number of repeated analyses

no distinction can be found between two increments provided $s_g < s_a$. (It is known a priori that an increment will not be equal to the object. This is not important, however, if the difference cannot be seen.)

The condition set for an increment to be a sample is $s_g \leqslant s_a$, or

$$100\left(\frac{P_A P_B}{n}\right)^{1/2} < s_a \qquad (2.3)$$

which is equivalent to

$$n > \frac{P_A P_B \times 10^4}{s_a^2} \qquad (2.4)$$

Other approaches are possible. Visman described the sampling variance as the sum of random and segregation compounds according to $S^2 = (A/wn) + (B/n)$. Here A and B are constants determined by preliminary measurements on the system. The A term is the random component; its magnitude depends on the weight w and number n of sample increments collected as well as the size and variability of the composition of the units in the population. The B term is the segregation component; its magnitude depends only on the number of increments collected and on the heterogeneity of the population. Different ways of estimating the values of A and B are possible.

Another way to estimate the amount of samples that should be taken in a given increment so as not to exceed a predetermined level of sampling uncertainty is via the use of Ingamells's sampling constant (IN 73, IN 74b, IN 76). Based on the knowledge that the between-sample standard deviation decreases as the sample size is increased, Ingamells has shown that the relation $WR^2 = K_s$ is valid in many situations. Here W represents the weight of the sample; R is the relative standard deviation (in percent) of the sample composition; and K_s is the sampling constant, the weight of sample required to limit the

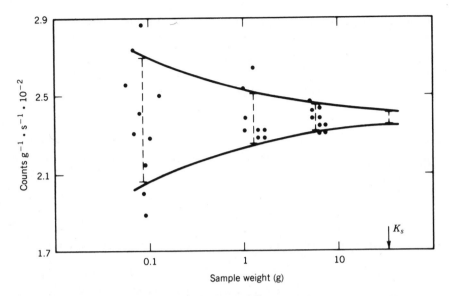

Figure 2.4. Sampling diagram of sodium-24 in human liver homogenate (HA 80). Reprinted courtesy of the National Institute of Standards and Technology, from NBSIR 80-2164, *Technical Activities 1980, Center for Analytical Chemistry*, edited by C. W. Reimann, R. A. Velapoldi, L. B. Hagan, and J. K. Taylor, 1980. Not copyrightable in the United States.

sampling uncertainty of 1% with 68% confidence. The magnitude of K_s may be determined by estimating the standard deviation from a series of measurements of samples of weight W. An example of an Ingamells sampling constant diagram is shown in Figure 2.4 (HA 80).

Gy (GY 82) collected the prevailing techniques in his book on sampling particulate materials and developed an alternative way of calculating the sample size. He defined a shape factor f as the ratio of the average volume of all particles having a maximum linear dimension equal to the mesh size of a sieve to that of a cube which will just pass the same screen: $f = 1.00$ for cubes and 0.524 for spheres; for most materials; $f \sim 0.5$. The particle size distribution factor g is the ratio of the upper size limit (95% pass screen) to the lower size limit (5% pass screen). For homogeneous particle sizes $g = 1.00$. The composition factor c is

$$c = \frac{(1 - x)[(1 - x)d_x + x d_g]}{x} \tag{2.5}$$

where x is the overall concentration of the component of interest; d_x is the density of this component and d_g the density of the matrix (remaining components); c is in the range $5 \times 10^{-5}\, kg/m^3$ for high concentrations of c to 10^3 for trace concentrations. The liberation factor l is

$$l = (d_l/d)^{1/2} \tag{2.6}$$

where d_l is the mean diameter of the component of interest, and d the diameter of the largest particles. The standard deviation of the sample s is estimated by

$$s^2 = fgcld^3/w \tag{2.7}$$

Ingamells (IN 74a) related Gy's sampling constant to Ingamells's constant by

$$K_s = fgcl(d^3 \times 10^4) \tag{2.8}$$

A computer program based on this method is described by Minkkinen (MI 87).

Brands proposed a calculation method in the case of segregation (BR 84). A special type of inhomogeneous, particulate objects is the surface analysis by microscopic techniques, e.g., analytical electron spectroscopy, laser-induced mass spectroscopy, or proton-induced X-ray emission. Here the minimum sample size can be translated into the minimum number of specific sample points in the specimen under investigation.

Morrison's group (FA 77, SC 77a) defined a sampling constant

$$K = \left(\frac{1}{I_T}\right)\left(A_T \sum_{i=1}^{n} i_i^2\right)^{1/2} (1 - A_I/A_T)^{-1/2} \tag{2.9}$$

where K = sampling constant in μm
  i_i = intensity of inclusion i
  n = total number of inclusions
  A_T = total area image in μm^2
  A_I = total area inclusions

For a confidence interval Δ the number of replicate analyses can be calculated

$$N = (100tK/\Delta^{1/2})^2 \qquad (2.10)$$

where a = area sampled

   t = Student's-t

Inczédy (IN 82) proposed a measure for homogeneity of samples, assuming a sinusoidal change of concentrations of the component of interest. The concentration difference Δc can be calculated by

$$\Delta c < 4.14s \frac{(\pi \delta z/\Delta z)}{\sin(\pi \delta z/\Delta z)} \qquad (2.11)$$

where s = measurement error

   $\delta z/\Delta z$ = relative "window" (see Figure 2.5)

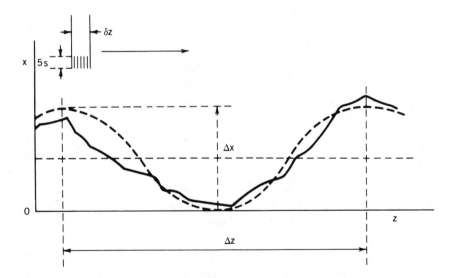

Figure 2.5. Concentration distribution of an element in a solid. The resolution of the instrument is high compared to the concentration change. The error is also very low ($\Delta z \gg \delta z$; $\Delta x \gg S$). The vertical lines indicate the uncertainty of the measurement; the standard deviation s can be taken as a fifth of the range of the signals for a given level (IN 82). Reprinted with permission from *Talanta* **29**, 643 (1982). Copyright 1982 Pergamon Press.

Danzer et al. (DA 79) proposed another homogeneity criterion based on Fisher's F-test for samples of different size.

2.6.1.3.   Internally Correlated Objects

Objects can be internally correlated in time or space; for example, the composition of a fluid emerging from a tank does not show random fluctuations but is correlated to the composition in earlier or later sampled product.

When the object is internally correlated, no distinct boundaries exist. In this case the sampling theorem of Nyquist can be used. This tells us that the sampling frequency must be twice as high as the highest frequency that occurs in the object. This rule works very well in physics, where many phenomena are expressed in waveforms that can be converted more of less easily into the frequencies by Fourier analysis. In chemistry, however, this approach is more difficult (Sections 4.3.2 and 4.3.2.2).

To know the composing frequencies, one must know the exact composition of the object with a very high resolution in order to construct the waveform. In this case it is easier as a rule to estimate the time constant or correlation constant derived from the autocorrelation function, for this parameter can be obtained in many ways, as given in Section 4.5.2. To obtain a high resolution, the "reconstruction" method that will be used for process control can be employed (Section 2.6.4).

Unmixed stratified objects such as soil and rock usually cannot be represented by a simple correlation constant derived from the exponential autocorrelation function. Here another measure is introduced, the semivariance.

As mentioned in Section 2.4, factors causing the internal correlation of objects include (1) diffusion or mixing within the object, for example, in mixing tanks, buffer hoppers, and rivers, and (2) varying properties of the producer of the object, reactors or emitters, for instance. In both situations samples are mutually dependent. When an object shows a large internal correlation, two adjacent samples do not differ much from each other. The difference between two samples increases with greater distance, however. One may ask how to estimate the number of increments that have to be taken to get a "real" sample or gross sample. As stated for the case of the heterogeneous objects, the number of increments n depends on the required variance of the sample, this variance being so small that the difference between two samples cannot be detected.

The sample size is influenced by many factors, including the lot size. Unfortunately, the equations that describe the variance of the sample cannot be simply altered to give the required number of increments but must be estimated by iterative methods. Another method is to derive them from a graphic representation of the equations. When the object to be analyzed is a process, a stream of material of infinite length with properties varying in time, the sampling parameters can be derived from the process parameters. When the object is a finite part of a process, however, usually called a lot, the description of the real composition of the lot depends not only on the parameters of the process the lot is derived from but also on the length of the lot. The sample now must represent the lot, not the process (Figure 2.2).

Lots derived from Gaussian, stationary, stochastic processes of the first order allow a theoretical approach. In practice most lots seem to fulfill the above requirements with sufficient accuracy to justify eqs. 2.12 and 2.13 (KA 78, Mu 78a): an estimate of the mean m of a lot with size P can be obtained by taking n samples of size G, equally spaced with a distance A; in this case $P = nA$ and the size of the gross sample $S = nG$. Here it is assumed that it is not permissible to have overlapping samples and that $A \rightarrow G$.

According to Müskens (MU 78b), the difference between the sample and the object or the lot the sample is taken from, expressed as the variance σ_*^2, can be thought of as composed of the variance in the composition of the sample σ_m^2, the variance in the composition of the whole object σ_μ^2, and the covariance between m and μ, $\sigma_{m\mu}$:

$$\sigma_*^2 = \sigma_m^2 + \sigma_\mu^2 - 2\sigma_{m\mu} \tag{2.12}$$

These variances can be calcuated with

$$\sigma_m^2 = \frac{2\sigma_x^2}{ng^2} \left\{ g - 1 + \exp(-g) + [\exp(-g) + \exp(g) - 2] \right.$$
$$\left. \cdot \left(\frac{\exp(-a)}{1 - \exp(-a)} - \frac{\exp(-a)[1 - \exp(-p)]}{n(1 - \exp(-a))^2} \right) \right\}$$

$$\sigma_\mu^2 = \frac{2\sigma_x^2}{p^2}[p - 1 + \exp(-p)]$$

(2.13)

$$\sigma_{m\mu}^2 = \frac{\sigma_x^2}{npg}\left\{2ng + [1 - \exp(-p)]\left[\frac{\exp(-g) - 1}{1 - \exp(-a)} + \frac{\exp(g) - 1}{1 - \exp(a)}\right]\right\}$$

where σ_x^2 = variance of process
$p = P/T_x$
$g = G/T_x$
$a = A/T_x$
T_x = correlation factor of the process
n = number of samples

The properties of the process from which the lot stems are described by σ_x, the standard deviation of the process, and T_x, the correlation factor of the process (Section 4.5.2).

The only relevant property of the lot here is its size p, expressed in units T_x. These units may be time units, such as when T_x is measured in hours. In this case T_x is usually called the time constant of the process. The lot size is expressed in hours as well. When a river is being described, the lot size can be the mass of water that flows by in 1 day or year.

The correlation factor T_x can be called a space constant, however, when the unit used is of length. Here the lot size is expressed in length units. This is the case, for example, when a lot of manufactured products contained in a conveyor belt or stored in a pile of material in a warehouse is considered. The values of P, G, and A can be expressed in dimensionless units when they are divided by T_x.

The correlation factor can also be expressed in dimensionless units, as when bags of products are produced. In this case the lot size is expressed in terms of items as well (number of bags, drums, tablets). The properties of the sample are increment size g, expressed in the same units as T_x and p, the distance between the middle of adjacent increments a and the number of increments n that form the sample. If the sample size is expressed as a fraction of the object $P(F = G/P)$, the relations between F and n are depicted in Figure 2.6 for a certain value of σ_x. The relationship of σ_*, F, and n for a number of values of p is depicted in Figure 2.7. The figures represent graphically the rather

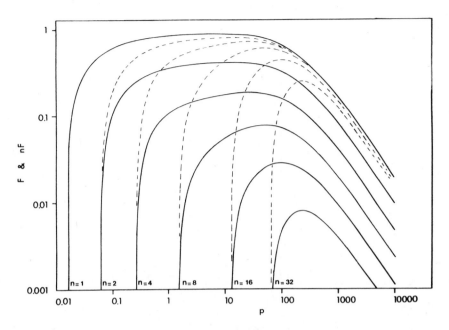

Figure 2.6. The relative sample size F (——) and the relative gross sample size nF (---) as a function of the relative lot size p for various numbers of samples ($\sigma_*/\sigma_x = 0.1$) (KA 78).

awkward equation 2.13 and can be used for setting up a sampling strategy.

For medium lot sizes P, or values of T_x between 0 and ∞, the smallest sample needed to obtain a certain value is one made up from many small increments. When fewer but larger increments are taken, the size of the sample increases. For $F = 0$, therefore, with infinite small increments and a given value of n, σ_*/σ_x can be equal to 0.1 only for values of P larger than those indicated in Table 2.3. If the lot size P is less than these values, σ_*/σ_x is less than 0.1. For larger values of P and σ_*/σ_x constant, log F is proportional to log P.

If the sample size is chosen such that the subsequent samples are taken without interruption, it can be shown that $\sigma_* = 0$. This result is not surprising because now the whole lot has been sampled: $A = P/n$, and the sample size $nG = P$. For an uncorrelated lot, with $T_x = 0$, it can be shown that σ_m, σ_μ, $\sigma_{m\mu}$, and consequently σ_* are all 0. If, however, the sample size $G = 0$, $\sigma_m^2 = \sigma_x^2/n$ and $\sigma_*^2 = \sigma_x^2/n$. In other words, if the lot size is very large in time constant units, one sample of finite size suffices,

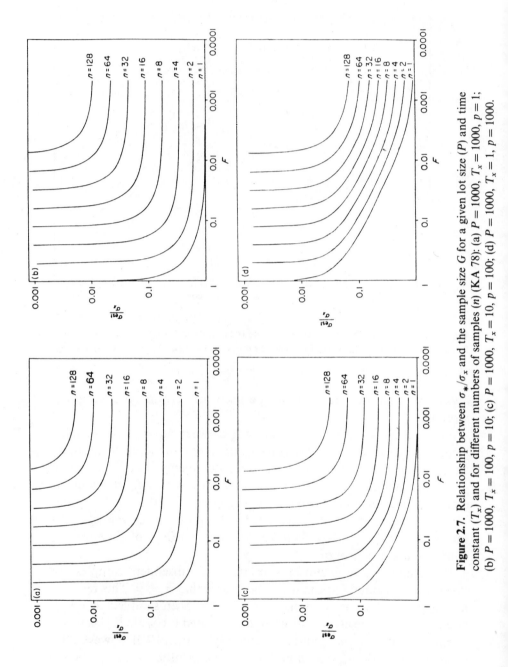

Figure 2.7. Relationship between σ_*/σ_x and the sample size G for a given lot size (P) and time constant (T_x) and for different numbers of samples (n) (KA 78): (a) $P = 1000$, $T_x = 1000$, $p = 1$; (b) $P = 1000$, $T_x = 100$, $p = 10$; (c) $P = 1000$, $T_x = 10$, $p = 100$; (d) $P = 1000$, $T_x = 1$, $p = 1000$.

Table 2.3. Minimal Values of p to Obtain $\sigma_*/\sigma_x = 0.1$ ($F = 0$)

n	1	2	4	8	16	32	64	100
p	0.0151	0.0614	0.2640	1.508	12.64	64.34	342.0	∞

Source: Reprinted with permission from G. Kateman and P. J. W. M. Müskens: *Anal. Chim. Acta* **103**, 11–20 (1978).

but if the increment size is near zero, a number of samples must be taken. Note that here the sample can be obtained from one part of the object (e.g., when $G \neq 0$) or from several, equally spaced places (e.g., when $G = 0$).

When $T_x \rightarrow \infty$ the situation of a homogeneous object is approached. In this case it can be shown that σ_m, σ_μ, and $\sigma_{m\mu}$ all approach the value of σ_x; therefore σ_* will be 0 and one sample of any desired size suffices. In Figure 2.6 the relation of P, F, and n for a constant value of $\sigma_*/\sigma_x = 0.1$ is given. At low values of P both a small increment size F and a small number of increments n are needed to obtain this value. This means that in highly autocorrelated lots only one small sample is needed to describe the lot with sufficient accuracy.

In Figure 2.7 the relation between n, G, and σ_*/σ_x for some values of T_x and a fixed value of P is given. It is possible to reach lower values of σ_*/σ_x for medium values of T_x in two ways: by increasing the number of samples n or by increasing the increment size G. Over certain values of G increasing n is impossible because nG, the size of the gross sample, cannot exceed the lot size P without overlapping increments G. For lower values of T_x an increase in sample size F causes a larger improvement in the precision than for higher values of T_x. Table 2.4 shows F, the increment size G as fraction of P, needed for one increment

Table 2.4. Comparison of Various nF Combinations That Yield the Same Ratio σ_*/σ_x

P	n	F	σ_*/σ_x	n	F
1000	32	0.01	0.060	1	0.350
100	32	0.01	0.079	1	0.730
10	32	0.01	0.031	1	0.960
1	32	0.01	0.015	1	0.980

Source: Reprinted with permission from G. Kateman and P. J. W. M. Müskens: *Anal. Chim. Acta* **103**, 11–20 (1978).

to reach the same precision as is reached by 32 samples of size $F = 0.01$. From this table and Figures 2.6 and 2.7 it can be concluded that for lots with $P < 1000$ it is better to use a large number of small increments than a small number of large increments. Other examples of the use of autocorrelation in sampling are given by Bobec (BO 73, BO 79) and Nelson (NE 81).

In soil science one usually has to deal with two- or three-dimensional objects that cannot be represented by correlation constants. Here the size and the number of samples must be obtained in other ways. In this case often an intermediate step is *kriging*, i.e., mapping of lines or planes of equal composition (WE 92). Kriging can be seen as an optimal unbiased interpolation method; based on known values in sampled locations, values at other locations are predicted. A measure indicating the amount of correlation between two locations separated τ times a lag distance h apart is the semivariance γ_τ. The semivariance for a lag τ is given by

$$\gamma_\tau = E\left\{[z(x) - z(x + \tau)]^2\right\} \tag{2.14}$$

and can be estimated by the formula:

$$\gamma_\tau = \frac{1}{2n(\tau)} \sum_{i=1}^{n(\tau)} [z(x_i) - z(x_i + \tau)]^2 \tag{2.15}$$

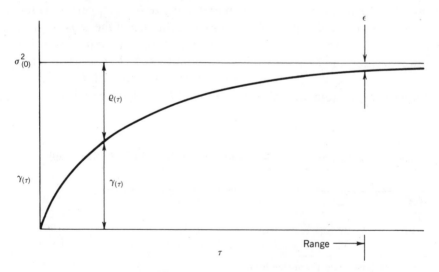

Figure 2.8. Semivariance.

where $z(x)$ and $z(x_i + \tau)$ the values of the variable of interest,[2] x, at location i and a location τ lags away, respectively. The number of pairs $[z(x_i), z(x_i + \tau)]$ is given by $n(\tau)$. The semivariogram, in which semivariances are plotted against the lag number (see Figure 2.8), is modeled by fitting a simple function through the data points. Linear, exponential, and spherical models are often used. At lag zero, the semivariance is 0. In many cases a discontinuity is present called the nugget effect. This indicates that there is significant variance over distances smaller than the lag distance. At large distances, the semi-variance is equal to the variable variance and reaches a ceiling, called the sill. These parameters, and the range where the semivariance increases with the lag, are indicated in Figure 2.8.

The value of a variable x at an unsampled location is derived from a linear combination of nearby points. Kriging attempts to minimize the prediction error. This can be achieved by solving matrix equation 2.16:

$$\begin{vmatrix} \gamma(h_{1,1}) & \cdots & \gamma(h_{1,n}) & 1 \\ \vdots & & \vdots & \vdots \\ \gamma(h_{1,n}) & \cdots & \gamma(h_{n,n}) & 1 \\ 1 & \cdot & 1 & 0 \end{vmatrix} \begin{vmatrix} W_1 \\ \vdots \\ W_n \\ \lambda \end{vmatrix} \begin{vmatrix} \gamma(h_{1,p}) \\ \vdots \\ \gamma(h_{n,p}) \\ 1 \end{vmatrix} \qquad (2.16)$$

or the equivalent form

$$A \cdot \mathbf{w} = \mathbf{b} \qquad (2.17)$$

where $h_{i,j}$ is the distance between the sampled locations i and j; $\gamma(h_{i,j})$ is the semivariance between locations i and j; $\gamma(h_{i,p})$ is the semivariance between the sampled location i and the unsampled location p, obtained from the semivariogram; W_i is the weight given to location i in the prediction of location p; and λ is the Lagrange multiplier. The solution for the unknown vector $\mathbf{w}$ is simple:

$$\mathbf{w} = A^{-1} \cdot \mathbf{b} \qquad (2.18)$$

[2] In kriging and related topics, variables are termed *regionalized*, indicating that the variables are a nonrandom sample from one realization of a random function. This is rather different from the usual idea of multiple realizations of a random variable. For the present purposes, this difference can be ignored.

For an estimate on location p, this gives the value

$$\hat{x}_p = \mathbf{w} \cdot \mathbf{x} \tag{2.19}$$

with prediction variance

$$s_\varepsilon^2 = \mathbf{w} \cdot \mathbf{b} \tag{2.20}$$

The weights obtained in this way are optimal in the sense that they are unbiased and have a minimum estimation variance. The predictor is also exact, which means that the value at a sampled location is predicted exactly and with zero error. Contour maps may be drawn that provide a global view of the situation.

Several other types of kriging exist that make a more exact prediction possible. In universal kriging (ST 91, WE 80), a polynomial trend is permitted. In co-kriging (MB 83, MY 82), the value of the variable of interest in an unsampled location is not only determined by interpolation from sampled locations but also by measurements of correlated variables. Disjunctive kriging (WE 89, YA 86a, YA 86b) is a variant where the conditional probability that a soil value exceeds a predefined threshold can be plotted.

By estimating "compartments" of compositions that do not vary more than a given amount, sampling can be restricted to one sample per compartment. Most compartment estimates are arbitrary, however (BO 83, BO 85, GR 85, WE 83).

2.6.1.4. Practical Applications (KA 78)

Sampling a Lot of Correlated Items. A certain bulk chemical is sold in lots of 10,000 bags of 25 kg each. By sampling the product stream during production, the autocovariance function (Section 4.5.2) was calculated and the parameters σ_x and T_x were estimated (Figure 2.9). Assume the product is sold under the condition that the analysis of the whole lot will not deviate by more than 0.3% from the specified composition. When a 95% confidence level is adopted, the standard deviation of the gross sample should not exceed 0.1%. Accounting for the standard deviation of the method of analysis (1.25%), the value of $\sigma_*/\sigma_x = 0.08$. The sampling scheme of Table 2.5 can be computed from eq. 2.13 or derived from an extension of Figure 2.7.

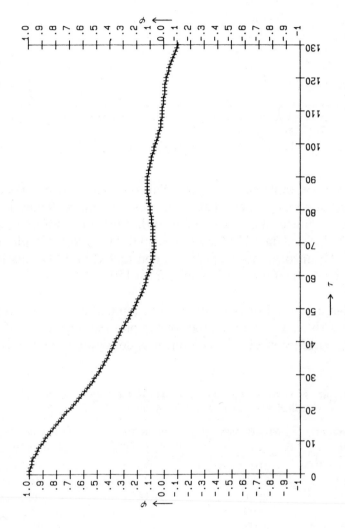

Figure 2.9. Autocovariance function of a product stream.

43

Table 2.5. Size of Gross Sample as a Function of the Number of Samples for a Specified Value of $\sigma_*/\sigma_x = 0.08^a$

No. of Samples (n)	Sample Size (F)	Sample Size (G = PF bags)	Gross Sample (nG bags)
1	0.91	9100	9100
10	0.045	450	4500
18	0.0014	14	452
19	0	1	19
20	0	1	20
40	0	1	40

Source: Reprinted with permission from G. Kateman and P. J. W. M. Müskens: *Anal. Chim. Acta* **103**, 11–20 (1978).

^{a} $\sigma_x = 1.25\%$; $\sigma_* = 0.1\%$; $T_x = 24{,}600 \text{ kg} = 984 \text{ bags}$; $P = 10{,}000 \text{ bags}$; $P = 10{,}000/984 = 10.2$.

Note that a sample size G of less than one bag still requires one sampling action. It is clear that, in this case, sampling 19 bags is the method of choice. When it is assumed that the lot is not internally correlated, the data of Table 2.6 are valid; here more samples are needed. The incorporation of the time constant T_x in the calculations can decrease the number of samples from 150 to 19.

Sampling a River. For estimation of the annual mean of a component in a river, a certain number of samples must be taken. If the internal correlation of the component is considered, the precision

Table 2.6. Size of Gross Sample as a Function of the Number of Samples for a Specified Value of $\sigma_*/\sigma_x = 0.08$, Assuming $T_x = 1$ (Bag)^a

No. of Samples (n)	Sample Size (F)	Sample Size (G = PF bags)	Gross Sample (nG bags)
1	0.03	300	300
10	0.0029	29	290
20	0.0014	14	280
50	0.00048	5	250
100	0.00014	2	200
150	0	1	150
200	0	1	200

Source: Reprinted with permission from G. Kateman and P. J. W. M. Müskens: *Anal. Chim. Acta* **103**, 11–20 (1978).

^{a} $\sigma_x = 1.25\%$; $\sigma_* = 0.1\%$; $P = 10{,}000 \text{ bags}$; $p = 10{,}000/1 = 10{,}000$.

Table 2.7. Parameters of the Rhine River for the Concentration and Load Values of Two Components, Estimated from Data Measured in the Period 1971–1975

Variable	$\bar{x}$ (mg/L)	σ_x (mg/L)	σ_a (mg/L)	T_x (days)
NH_4^+				
Concentration	3.07	1.87	0.62	120
Load	4.91	2.07	0.93	50
NO_3^-				
Concentration	13.43	4.79	1.47	90
Load	24.22	11.71	2.67	30

Source: Reprinted with permission from G. Kateman and P. J. W. M. Müskens: *Anal. Chim. Acta* **103**, 11–20 (1978).

will be higher. Alternatively, an accepted precision can be obtained with fewer samples. For the concentration and load of some components of the Rhine River at Bimmen (Germany), T_x was estimated from the autocovariance function (Table 2.7). Here σ_x is the standard deviation of the composition; σ_a, the standard deviation of the analysis.

The error of the determination of the mean value of a component during 1 year, Δ_*, depends on the values of σ_* and σ_a, the precision of the method of analysis. If we assume that each sample is followed by an analysis, the ultimate precision is $\sigma_L^2 = \sigma_*^2 + \sigma_a^2/n$. For a confidence level of 0.05, $\Delta_* = t_n\sigma_L$, where t_n is Student's-t for a confidence level of 0.05 and n determinations. The calculations are performed for two samples sizes, $G = 0$ (in this case the samples are called momentary) and $G = A$, if the samples are integrated between the sampling actions. In the latter case, with $nG = P$, the value $\Delta_* = t_n\sigma_a/n^{1/2}$ is to be expected. For both cases the relationship between Δ_* and the sampling interval A or the number of samples n, calculated for the concentration of ammonia in the Rhine River, are shown in Figure 2.10. The same relationship is also shown for $G = 0$, neglecting the autocorrelation ($T_x = 0$). If the number of samples n is larger than 130, the results for $G = 0$ and $G = A$ are practically identical. Furthermore, it can be shown that samples collected over 1 day ($G = 1$) barely give a better estimation of the annual average than the momentary samples ($G = 0$).

From Figure 2.10 conclusions can be drawn about the sampling strategy. The best strategy (with the lowest value of Δ_*) is to take a

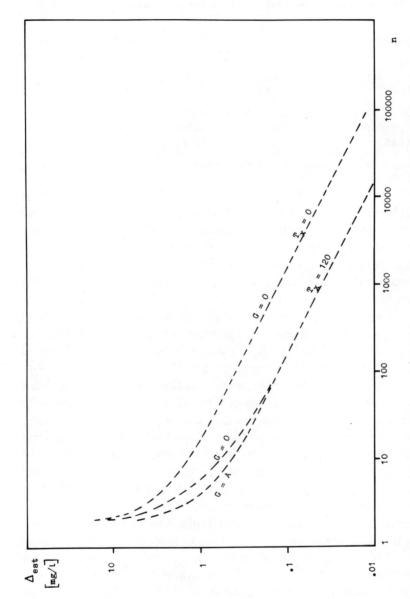

Figure 2.10. The error in the estimation of the annual average Δ_*, as a function of the number of samples n (KA 78).

46

sample collected over the whole period P, which should be analyzed afterward by a very precise method. However, if a precise method is not available or if the samples cannot be conserved over the whole period, more daily samples must be taken and analyzed. The number of samples needed to obtain a certain value of Δ_* can be deduced from Figure 2.10.

Frequently, a relative error $\Delta_*/\bar{x}$, where $\bar{x}$ is the mean value of the variable, is required. Table 2.8 shows how many samples are needed to obtain the ratio $\Delta_*/\bar{x} = 0.1, 0.5, 0.01,$ or 0.005. These calculations

Table 2.8. Number of Samples n Needed to Obtain a Certain Ratio $\Delta_*/\bar{x}$ in Estimating the Annual Average of a Variable Investigated in the Rhine River[a]

Variable	Ratio $\Delta_*/\bar{x}$			
	0.1	0.05	0.01	0.005
NH_4^+				
Concentration	2	73	1600	6300
Load	22	65	1400	5500
NO_3^-				
Concentration	12	29	480	1900
Load	19	43	510	1900

Source: Reprinted with permission from G. Kateman and P. J. W. M. Müskens: *Anal. Chim. Acta* **103**, 11–20 (1978).

[a] $G = 0$; other data in Table 2.7.

Table 2.9. Number of Samples n Needed to Obtain a Certain Ratio $\Delta_*/\bar{x}$ in Estimating the Annual Average of a Variable Investigated in the Rhine River, If Autocorrelation Is Neglected[a]

Variable	Ratio $\Delta_*/\bar{x}$			
	0.1	0.05	0.01	0.005
NH_4^+				
Concentration	160	640	16000	63000
Load	85	330	8200	33000
NO_3^-				
Concentration	56	220	5400	21000
Load	97	380	9500	38000

Source: Reprinted with permission from G. Kateman and P. J. W. M. Müskens: *Anal. Chim. Acta* **103**, 11–20 (1978).

[a] $T_x = 0$; $G = 0$; other data in Table 2.7.

were performed for the estimation of the annual average of the variables of Table 2.7. The values of n, assuming no internal correlation ($T_x = 0$), are given in Table 2.9.

2.6.2. Samples for Description in Detail

In contrast with the gross description of objects, where one parameter, the mean, suffices to describe the object, here the composition as function of time or space is wanted. Such a situation is met in metallurgical analysis, for example, when the compositions of the various phases are to be determined.

One may ask how many samples are needed to reconstruct the object in sufficient detail. Every sample must fulfill the condition that it resembles the object, that is, the part of the object that has to be reconstructed, as well as possible by the method of analysis and the details sought.

2.6.2.1. Heterogeneous Objects

If we assume that the object to be described consists of many objects, each homogeneous and randomly distributed, it is clear that the sample size must be smaller than or equal to that of the object. A restriction is that the method of analysis must be able to discriminate between the composition of the constituting objects. The number of samples cannot be estimated, for this is set by the following problem: is the object to be known exactly in every detail, or will it be sufficient to know the structure of the object? In this case the number of samples is given by the rules of Section 2.6.1.

2.6.3. Samples for Threshold Control

Processes are often monitored to report the day-to-day or hour-to-hour state of the process. The purpose of monitoring can be detection of a predetermined threshold value, therefore raising an alarm. There are numerous examples of such procedures in practice, such as detecting an explosion limit and seeking the maximum allowable concentration of a catalyst poison or of a pollutant in an effluent.

When the only function of the monitoring system is to raise an alarm, many samples are superfluous because as a rule the process value does not exceed the threshold levels. An optimal situation may be defined to be where the number of samples is such that a predetermined fraction of all threshold transgressions is detected. The selected value of this fraction depends on the costs (expressed in some measure, e.g., money) of effecting sampling and measuring, and the costs that accompany exceeding the threshold level.

Two ways to tackle this problem are described. The first is to adapt the sampling frequency to the last known process value: if this value is remote from the threshold level, the sampling frequency is low. If, however, the process value approaches the threshold level, the sampling frequency is enhanced. The second approach is to fix the sampling frequency according to the mean probability of threshold transgressions. Which way is preferred depends on the method of sampling and the absolute values of the times between samplings. In some instances a rigid sampling scheme is required; in other instances, for example, in sampling a river with a large time constant, sampling may depend on the results of the last sample.

2.6.3.1. *Variable Distance Between Samples*

When monitoring is performed with the aim of detecting a transgression of an upper or lower threshold level, or both, the most reliable scheme is to measure continuously, for in that way all threshold transgressions are detected.

As a rule not as many analyses are required when not all transgressions have to be detected. When the cost of analysis exceeds the cost due to excessive process values, such a situation becomes interesting.

When the situation of a first-order stochastic process is assumed, which is often encountered, the frequency of sampling can be adapted to the process situation as described by Müskens (MU 78b).

Let us assume that the process to be monitored has a time constant T_x, as follows from the autocorrelation (Section 4.5.2), a mean process value μ_x, and a standard deviation of the process σ_x. The threshold level x_g can be chosen or will be dictated by the environment.

Future values can be predicted according to the equation

$$\hat{x}_{t+\tau} - \mu_x = \alpha^\tau (x_t - \mu_x) \tag{2.21}$$

where $\hat{x}_{t+\tau}$ = predicted value of x, τ units of time after now (t)

x_t = process value at time t

The variance of the prediction is expressed by

$$\hat{\sigma}_\tau^2 = \sigma_x^2(1 - \alpha^{2\tau}) \tag{2.22}$$

For first-order stationary stochastic processes $a = \exp(-1/T_x)$. Thus after each measurement x, the probable future values of x, when predicted over a time span τ, are found in a region with the boundaries

$$\text{upper} = \hat{x}_{t+\tau} + Z\hat{\sigma}_\tau = \alpha^\tau(x_t - \mu_x) + \mu_x + Z\sigma_x(1 - \alpha^{2\tau})^{1/2} \tag{2.23}$$

and

$$\text{lower} = \hat{x}_{t+\tau} - Z\hat{\sigma}_\tau = \alpha^\tau(x_t - \mu_x) + \mu_x - Z\sigma_x(1 - \alpha^{2\tau})^{1/2} \tag{2.24}$$

where Z is the reliability factor that sets the probability p that the real process value $x_{t+\tau}$ will be within the reliability interval. This factor Z can be chosen.

As long as the upper limit of the reliability interval does not cross level x_g (Figure 2.11), the probability that the threshold level will be

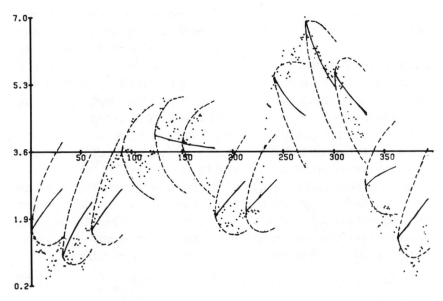

Figure 2.11. Operation of the monitoring system (MU 78b): ($\cdots$) measurements; (—) prediction; (---) 95% reliability interval of the prediction.

exceeded is less than $1 - p$ and sampling is not necessary. However, as soon as the upper level reaches x_g, sampling should be started. From the result thus obtained the time for the next sampling may be derived in the same way. By application of this procedure, which is based on the most recent information, the sampling frequency will be lowered automatically if the process values are remote from the threshold x_g. The frequency increases when the process values approach x_g. The prediction period τ, during which time no samples have to be taken, can be derived by fixing $x_g = $ upper (or lower) limit.

Then

$$\tau = -T_x \ln \left\{ \frac{T_r \cdot X_t + Z[X_t^2 - q(T_r^2 - Z^2)]^{1/2}}{X_t^2 + qZ^2} \right\} \qquad (2.25)$$

where $T_r = $ threshold value (normalized to zero mean and unit standard deviation)

$X_t = $ process value (normalized to zero mean and unit standard deviation)

$Z = $ reliability factor

$q = (\sigma_x^2 + \sigma_a^2)/\sigma_x^2$

$\sigma_x^2 = $ variance process

$\sigma_a^2 = $ variance method of analysis

Note that Z must be chosen greater than T_r.

The relationship between τ/T_x, the period where the prediction is reliable and sampling is not necessary, and the value of the last measurement x_t is shown in Figure 2.12. Suppose a process with a time constant $T_x = 100$ is at its set point, so the normalized process value $x_t = 0$. We want to know the future period τ when the process value will be within the boundaries chosen. [Suppose $T_r = 1$ (SD) with a reliability $Z = 2$ ($\sim 95\%$; Z is comparable to Student's).] Equation 2.25 is used to calculate τ, but the value can also be found in Figure 2.12. Under the chosen circumstances this value is 14. When we increase the threshold value T_r to 1.5, the next measurement can be postponed 41 time units. When the process value is higher, say 0.5, the prediction period τ is much shorter (about 3) because the process value is nearer to the safety boundary we have chosen.

The sampling frequency is fixed by the process parameters, the threshold value, and the reliability factor. Because the process param-

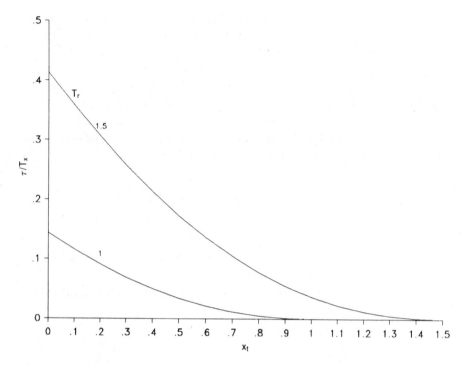

Figure 2.12. Relation between the predicted safe period without sampling τ and the value of the last measurement x_t for two threshold values T_r.

eters are inherent to the process and when the threshold level is set by the environmental conditions, the only degree of freedom is the reliability factor Z. The higher Z, the lower the probability of undetected threshold transgressions but the higher the sampling frequency. The optimal value of the reliability factor can be determined by balancing the costs of the analysis against the costs caused by undetected threshold transgressions.

As in practice it is difficult to have changing sampling intervals, the mean frequency of sampling $\bar{\tau}^{-1}$ can be calculated (MU 78b). Now the possibility that a threshold transgression takes place unnoticed will be larger. When the reliability factor R is defined as the fraction of detected transgressions of the total number of possible transgressions, some rules of thumb can be derived from the calculations of

Müskens:

$$Z > T_r + 0.5, \qquad R > 90\%$$
$$Z > T_r + 1.0, \qquad R > 98\%$$
$$Z > T_r + 1.5, \qquad R > 99\%$$

As previously indicated, the only degree of freedom of the monitoring system is the reliability factor Z. The reliability R increases with Z, and consequently the total cost caused by undetected threshold transgressions decrease. On the other hand, the sampling frequency increases with Z, and therefore the cost of the analysis increases. The optimal value of Z is obtained if the total costs are minimal. A rather elaborate method of finding this optimum is to compare the frequency of analysis with the reliability for corresponding values of Z and calculate the costs of both components. This should be repeated until the total minimal costs are found. A helpful way to find the optimum as quickly as possible is the simplex optimization method (see Section 4.9).

Example. The surveillance of the quality of surface waters presents a problem that is very suitable for practical application of the monitoring system. Müskens (MU 78b) described an application in this field, namely, the surveillance of the concentrations of ammonium, nitrite, and nitrate ions in the Rhine River.

The statistical properties of the time-dependent variations in the concentrations were estimated from daily measurements during the period 1971–1975. Here the total standard deviation s_y is a combination of the process standard deviation s_x and the standard deviation s_a of the analytical method. The time constant T_x was estimated from the autocorrelogram of the measured concentration values. Tests on normality for the data indicated that the concentrations were not normally distributed; therefore, a transformation was applied to the logarithms (Section 4.1.2).

The monitoring system was applied to the logarithms for a normalized threshold value $T_r = 2$ with reliability factors $Z = 3$ and $Z = 4$. The predicted mean interanalysis times were compared with the actual situation that was known because of the daily analysis. These values agreed very well. The following was the mean interanalysis time for ammonium ion:

$$T_r = 2 \qquad Z = 3 \qquad 4.8 \text{ days}$$
$$Z = 4 \qquad 3.1 \text{ days}$$

for nitrite ion:

$$T_r = 2 \qquad Z = 3 \qquad 3.0 \text{ days}$$
$$Z = 4 \qquad 2.0 \text{ days}$$

and for nitrate ion:

$$T_r = 2 \qquad Z = 3 \qquad 2.7 \text{ days}$$
$$Z = 4 \qquad 1.8 \text{ days}$$

Because the accepted frequency of analysis was 1 per day, it was concluded that considerable savings were possible if the proposed monitoring system was adhered to.

2.6.3.2. Fixed Distance Between Samples

When a varying sampling frequency is not appropriate, a fixed frequency can be chosen in such a way that the sum of control costs and costs caused by exceeding the threshold is a minimum (KO 71). The mean time between transgressions of the threshold limit, or the mean undisturbed run length T_r, can be determined by counting the number of transgressions m during a time T and defining $T_r = T/m$. The aim is

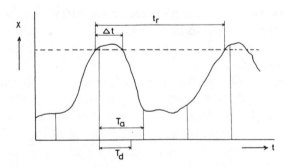

Figure 2.13. Process with undisturbed runlength t_r, transgression length Δt, distance between samples T_a, and delay time of analysis T_d (KO 71).

to detect transgressions of x_g, but because the only way to detect all transgressions is to sample continuously, an optimal value of the sampling frequency is sought.

Let us suppose the mean time between samplings is T_a (a sampling frequency of $1/T_a$) and the mean delay time (time required for sampling and analysis) is T_d (Figure 2.13). Because now the mean time between transgression detections is $\frac{1}{2}T_a$ and the transgression is not under control before action has been undertaken, the mean time of transgression ΔT is $\frac{1}{2}T_a + T_d$. (We assume that the effect of the control action is immediate, for very extreme control actions are possible, such as shutting off the endangering process stream.) The fraction of time the process suffers transgression is

$$\frac{\Delta T}{T_r} = \frac{\frac{1}{2}T_a + T_d}{T_r} \qquad \left(\frac{\Delta T}{T_r} \ll 1\right) \tag{2.26}$$

If the cost of transgression (losses of the production process, fines for elevated emissions, and costs of extra actions to prevent damage) is C_p ($/hour), the cost due to off-limit situations is

$$C_L = \frac{\frac{1}{2}T_a + T_d}{T_r} C_p \tag{2.27}$$

If the sampling cost (including sampling and analysis) is c_a ($/sample), the hourly cost is $C_a = c_a/T_a$. The total cost of control ($/hour) is the sum of C_a and C_L, or

$$C_t = \frac{C_a}{T_a} + \frac{\frac{1}{2}T_a + T_d}{T_r} C_p \tag{2.28}$$

For given or estimated values of c_a, C_p, T_d, and T_r, C_t is a function of T_a. This implies a minimum value of C_t when

$$(T_a)_{\text{opt}} = \left(\frac{2C_a T_r}{C_p}\right)^{1/2} \tag{2.29}$$

and

$$(C_t)_{\text{min}} = \frac{2C_a}{(T_a)_{\text{opt}}}\left[\frac{T_d}{(T_a)_{\text{opt}}} + 1\right] \tag{2.30}$$

As a rule estimates of C_p are difficult to obtain.

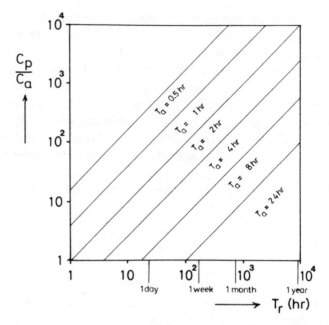

Figure 2.14. Nomogram for estimation of the optimum sampling frequency (KO 71).

Example. For use in practice it must be taken into account that $(T_a)_{opt}$ is reduced to "easy" values, for a sampling time of 2.735 hours, for example, would be very impracticable. Because plant laboratories generally operate at schedules of 0.1, 1, 2, 4, or 8 hours, following a system of three shifts a day, an effective sampling scheme must follow this schedule. A graph (Figure 2.14) can be produced that is based on these assumptions and uses only rough approximations of C_p/C_a.

2.6.4. Samples for Control

2.6.4.1. Parameters for Control[3]

Because the purpose of control is to diminish the variance of the process, the parameter that describes the success or the maximum available success of controlling action is based on diminution of the

---

[3] See also Section 3.2.5.

variance. The controllability factor r is defined as

$$r^2 = \frac{\sigma_x^2 - \sigma_{min}^2}{\sigma_x^2} = 1 - \left(\frac{\sigma_{min}}{\sigma_x}\right)^2 \qquad (2.31)$$

In this equation σ_x^2 = variance of process deviations without control and σ_{min}^2 = variance of process deviations after control. A factor $r = 1$ means that the variations in the process are fully eliminated and control is 100% successful. A factor $r = 0$ means that no diminution at all has occurred.

The controllability is determined by several factors, some related to the control mechanism, others related to the measuring device applied, and all related to the process to be controlled. According to van der Grinten (GR 65, GR 66), the controllability factor r can be divided into two parts, one related to control (r_p) and the other to measurement (m):

$$r = r_p m \qquad (2.32)$$

For the purpose of sampling, only m is of interest. The measurability factor m describes the influence of the measuring device on control. When the control action is perfect, $r_p = 1$; in such a situation it can be shown that m determines the maximum obtainable value of r:

$$m = m_D \cdot m_A \cdot m_G \cdot m_N \qquad (2.33)$$

where $m_D = \exp(-d)$ (measurability caused by delay time of analysis)

$m_A = \exp(-a/2)$ (measurability caused by sampling frequency A^{-1})

$m_G = \exp(-g/3)$ (measurability caused by sample size)

$m_N = 1 - s_a t_e^{1/2}$ (measurability caused by precision of analysis)

or

$$m = [\exp - (d + \tfrac{1}{2}a + \tfrac{1}{3}g)](1 - sa.t_e^{1/2}) \qquad (2.34)$$

where $d = D/T_x$, $a = A/T_x$, $g = G/T_x$, $t_e = T_e/T_x$, and $s = \sigma_a/\sigma_x$

D = analysis time

$A = $ (sampling frequency)$^{-1}$
$G = $ sample size
$T_x = $ time constant (correlation factor) of process
$T_e = $ time constant of measuring device
$\sigma_a^2 = $ variance of method of analysis
$\sigma_x^2 = $ variance of process

From eq. 2.34 it follows that the sample size should be kept to a minimum. In order to describe the influence of sampling it can be stated that

$$r = (m_D m_N) m_S r_P \qquad (2.35)$$

with m_S being the factor describing the sampling action

$$m_S = m_A m_G \qquad (2.36)$$

2.6.4.2. Reconstruction

Application of reconstruction methods can be described as the interpolation of object composition between sample points by means of an exponential function, characterized by the correlation constant (Figure 2.15).

2.6.4.3. Distance Between Samples

From eq. 2.33 it follows that the maximum obtainable controllability factor will never exceed the smallest of the composing factors. This implies that all factors should be considered in order to eliminate the restricting one. It also means that a trade-off is possible between high and low values of the various factors. If, for example, m_N, the factor influenced by the reproducibility of the measuring device or the method of analysis, is high and m_A is low, the frequency of sampling can be increased, causing m_A to increase. The reproducibility of the analysis can be decreased—which may result in a less expensive means of analysis—in order to cope with the higher rate of sampling if the product $m_A m_N$ is higher than before.

From $m_A = \exp(-a)$ it follows that decreasing a, the distance between samples or the reciprocal of the sampling frequency, causes m_A to

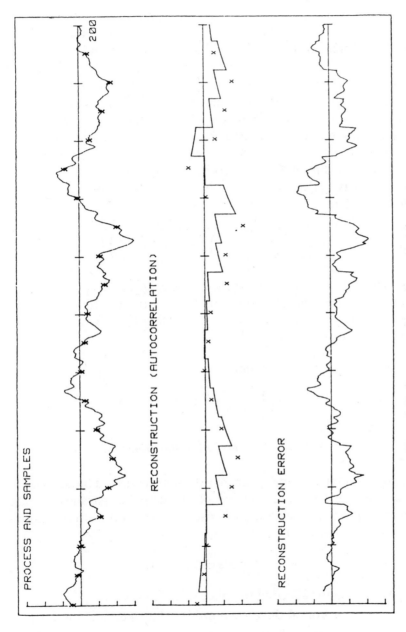

Figure 2.15. Reconstruction and reconstruction error.

59

increase. The higher the sampling rate, the better the possibility of controlling the process. This does not mean, however, that the highest obtainable frequency is the best, for sampling costs rise about linearly with $1/a$. Moreover, the optimal value of a depends on balancing the costs of analysis against costs of an eventual process failure, as is shown below. From $m_G = \exp(-g/3)$ it is clear that decreasing g, the sample size, increases m_G. The result is that, when sampling for control, it is always better to use samples that have the minimum size required by analysis. This is contrary to the situation when a lot is being described. A general statement can be made: never use integrating devices that collect samples during the time between analyses. If, for example, $g = a$, the sample is collected during the time between two consecutive samplings. Here

$$m = \exp\left[-\left(\frac{a}{2} + \frac{g}{3}\right) \right] = \exp\left[-\left(\frac{5a}{6}\right) \right] \qquad (2.37)$$

and high values of a result in low values of m. A special case is encountered when the time constant of the instrument, for example, an automatic analyzer, sets the sampling rate. Here $T_e = a$, so a third factor is influenced by sampling:

$$m_S = m_A m_G m_N$$
$$= \exp\left[-\left(\frac{a}{2} + \frac{g}{3}\right) \right](1 - s_a\sqrt{a}) \qquad (2.38)$$

2.6.4.4.   Practical Applications

To find an optimal sampling scheme, the scheme should be adjusted to the production cost as a function of the sampling parameters. The assumption that production costs rise exponentially with deviations from the set point yields

$$K' = [1 + \alpha(\exp|s| - 1)]K \qquad (2.39)$$

with α = factor affecting the price of deviations
$\quad\quad K$ = cost of product when produced at specification level
$\quad\quad s$ = dimensionless deviation from specification level expressed in SDs

The probability of producing at specification is

$$P = \frac{2}{(2\pi)^{1/2}} \exp\left(-\frac{1}{2}s^2\right) \tag{2.40}$$

The cost caused by deviations from the set point is the sum or integral of the deviations times the cost for each deviation:

$$k = K\left[1 + \frac{\alpha}{(2\pi)^{1/2}} \int_{-\infty}^{\infty} (\exp|s| - 1)\exp\left(\frac{-s^2}{2}\right)ds\right] \tag{2.41}$$

By control the deviations of the process are diminished and the expression becomes

$$k' = K\left[1 + \frac{\alpha}{(2\pi)^{1/2}} \int_{-\infty}^{\infty} (\exp|s| - 1)\exp\left(\frac{-s^2}{2(1-m^2)}\right)ds\right] \tag{2.42}$$

The influence of control on process costs can be expressed by

$$c = \frac{k'}{k} = \frac{\int_{-\infty}^{\infty} (\exp|s| - 1)\exp[-s^2/2(\cdots 1 - m^2)]\,ds}{\int_{-\infty}^{\infty} (\exp|s| - 1)\exp(-\frac{1}{2}s^2)\,ds} \tag{2.43}$$

When however the assumption is that production costs rise linearly, then

$$K' = (1 + \alpha|s|)K \tag{2.44}$$

and

$$c = \frac{\int_{-\infty}^{\infty} |s|\exp[-s^2/2(1-m^2)]\,ds}{\int_{-\infty}^{\infty} |s|\exp(-s^2/2)\,ds} \tag{2.45}$$

When the cost in arbitrary units is plotted as a function of precision and analysis frequency, as shown in Figure 2.16, it is clear that high

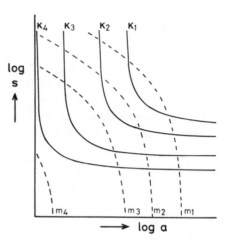

Figure 2.16. Cost and measurability as a function of precision (S) and analysis frequency (a^{-1}) [Reprinted with permission from *Chem. Weekbl.* May, p. 12 (1978)]. Cost: K_1, 100 arbitrary units; K_2, 200 arbitrary units; K_3, 500 arbitrary units; K_4, 1000 arbitrary units. Measurability: $m_1 = 0.8$; $m_2 = 0.9$; $m_3 = 0.95$; $m_4 = 0.99$.

precision or high sampling frequency are always expensive. The optimal combination should be sought for each process.

In order to provide some feeling for the cost factors under consideration the reader is referred to Leemans (LE 71), who published some data on the measurability and costs of an industrial process (Table 3.5 and Figure 2.17). The relation between m and cost is plotted in Figure 2.18.

In Figure 2.19 the cost of analyzing for control of a process and the measurability are related. The dashed line shows the process costs as a function of measurability, and it is clear that analysis costs should be lower. There exists a restricted area that sets boundaries to the variability of m. High values of m result in excessively increasing costs of analysis, unless the methods are fully automated. Low values of m are reflected in an undesired rise of the process costs. Note that Leemans uses a somewhat refined model for process costs. In his model he assumes that the margin left for process fluctuations in the guarantee conditions could be used if m is very high. This assumption implies that the extra process costs become negative at $m = 0.92$, as indicated in Figure 2.19.

A totally different approach that requires quite a lot of mathematics has been proposed by Ivaska (IV 86) but will not be treated here.

The sampling rate and therefore the information yield is not only set by the sampling scheme but also by the laboratory. If, for instance, the frequency of sampling is too high, queuing of samples occurs, which means loss of information if the results are used for process control.

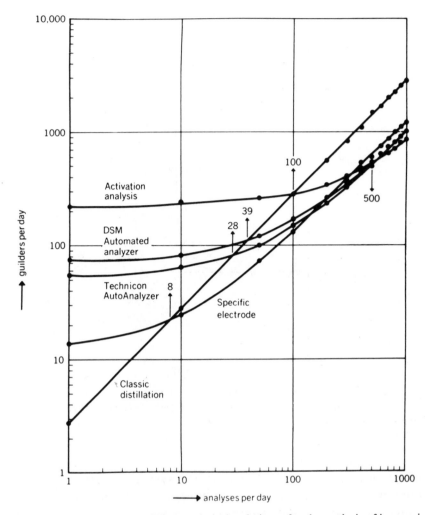

Figure 2.17. Costs of some off-line analytical techniques for the analysis of inorganic nitrogen (1968) (LE 71). Reprinted with permission from *Anal. Chem.* **43**(11), 46A (1971). Copyright 1971 American Chemical Society.

Janse (JA 83) showed that the theory of queuing can be applied to study the effects on information yield of such limited facilities. The effect depends on the way the sampling and analysis are organized.

The effect can be calculated for simple systems with one service point (one analyst or one instrument). For more complicated systems, simu-

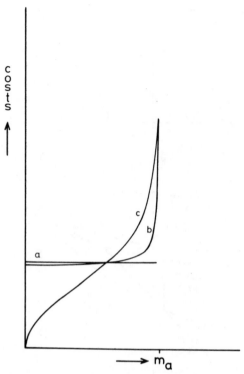

Figure 2.18. Cost of analysis as a function of measurability caused by analysis frequency [Reprinted with permission from *Chem. Weekbl.* May, p. 12 (1978)]. (a) capital intensive; (b) intermediate; (c) labor intensive.

lation should be applied. For a M/M/1 system[4] (random sampling/ random analysis time/one serving point), in fact the most unorganized system, the measurability factor is

$$m^2 = \frac{T_x}{T_x + 2\bar{A}} \cdot \frac{(1-\rho)\,T_x}{(1-\rho)\,T_x + \bar{D}} \qquad (2.46)$$

[4] In an A/B/C system, A depicts the type of arrival process, B depicts the type of service times, and C depicts the number of service channels (analysts). Here M means Markovian—each event is independent of the others; D means deterministic—each event is predictable.

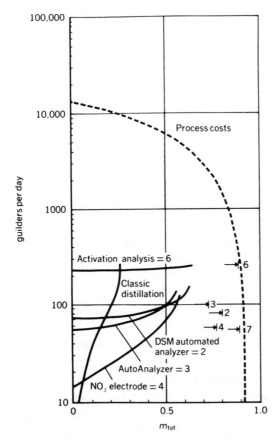

Figure 2.19. Process costs due to process fluctuations (dashed line) and analysis costs (solid line) as a function of analytical information. The arrows refer to the fully automated techniques (LE 71). Reprinted with permission from *Anal. Chem.* **43**(11), 44A (1971). Copyright 1971 American Chemical Society.

where T_x = time constant (correlation constant)

   $\bar{A}$ = mean interarrival time of samples

   $\bar{D}$ = mean interanalysis time

   ρ = utilization factor of the lab

For a D/D/1 system (fixed sampling rate/fixed analysis time/one serving point) three are no waiting times.

The situations D/M/1 (fixed sampling rate) and M/D/1 (fixed analysis time) are intermediate. Thus we have—

D/M/1 system:

$$m^2 = \exp\left(\frac{-2\bar{A}}{T_x}\right) \cdot \frac{(1-\rho)\,T_x}{(1-\rho)\,t_x + 2\bar{D}} \tag{2.47}$$

M/D/1 system:

$$m^2 = \frac{T_x}{T_x + 2\bar{A}} \cdot \frac{2(1-\rho)\bar{A}}{(2\bar{A} - T_x)\exp(2\bar{D}/T_x) + T_x} \tag{2.48}$$

One obvious conclusion can be that smoothing the input (from M/M/1 to D/M/1) is most advantageous. (See also Section 5.1.1.3 and Figures 5.9–5.11.)

2.6.5. Samples for Special Purposes

2.6.5.1. One-Shot Sampling

In many instances it is not possible to use any of the sampling strategies that are mentioned in the preceding sections. Such a situation is met, for instance, when sampling the moon, the remains of a burnt building, or evidence at the scene of a crime. This is so-called one-shot sampling. Only one sample can be obtained, of a size that cannot be related to the object to be described. Sampling the moon cannot be done by an astronaut or an automatic moon sampler (EN 73) in the way described in Sections 2.6.1 or 2.6.2. When describing the place to be sampled there is very little known about the distribution of the components, their composition, or the correlation, for example, so it is not possible to state the size or number of grabs that must be taken to constitute a real sample, representative of the entire moon. These circumstances do not occur solely on the moon but also when, for example, collecting geological samples on earth or taking samples in shops for the food inspection department. To overcome these problems it is possible to work the other way round. The size of the increment is determined by maximum weight, volume, or cost. The increment is stated to be a sample; when the sample has been analyzed, the parameters found are used to describe the lot size. In some situations the lot size is barely larger than the sample, such as when sampling parts of a large lot, say, bread in shops. Each loaf of bread sampled at the factory can be considered as a part of the bread-making process and will show correlation with loaves produced at the same time or found in the same oven. When the loaves are distributed in town, however, nothing can be said about the autocorrelation. The sample of one loaf equals a lot of one loaf.

When the moon is sampled, the sample brought back to earth should be allocated to a type of rock known to exist in a known area. Here the description of the lot by the sample is part of the job. Another example of one-shot sampling is examination of evidence at the scene of an accident. The piece of glass found near the victim is known to be part of a lot, from the windshield of a car. Here, too, the problem is to allocate the sample to a lot. There are two ways of doing so. One is to allocate the sample to an unspecified lot; for example, this glass is used in windshields of manufacturer X. The other method is to allocate it to a specified lot; for example, this is glass from the broken windshield of car Y.

In the case of one-shot sampling, care must be taken not to over-estimate the evidence provided by the sample. As was shown in the preceding sections, many properties of the lot or process must be known in order to ascertain the size of a sample that is representative of the lot. Only when these properties can be obtained from the sample itself is it justified to describe the lot using the data the sample yields.

2.6.5.2. Selected Sampling

Although it is good practice to define the sample as representing the lot or the process, there are situations in which this rule should not be applied. Two possibilities are described:

1. *Tracing deviations from the normal situation.* This occurs when faults in material must be found, mistakes, or forgeries, for example. Now the purpose is not to describe the whole process or lot with one sample but to prove that part of the process or lot is deviating. In order to relate the results derived from the sample to the part considered, this part can be taken as a partial lot and the normal rules can be applied. Note that a part of a process, smaller than that needed for an ergodic

part, by definition is a lot (see Section 2.6). Here the rules for lots are to be used.

2. *When deviations of the normal situation are suspected.* In many cases deviations are suspected, for example, when malfunction of the production process is possible or when forgeries are suspected. The recommendation in such a situation is to forget everything you know about sampling and use your common sense and experience. As soon as the deviation has been found, return to the scientific practice and use the rules of part 1, above.

2.7. HANDLING OF SAMPLES

The foregoing sections have been devoted to the theoretical requirements of sampling. In practice, however, the value of a sample is determined not only by the prerequisites of time and size but also by the handling of the sample. This means that the sample must be taken, homogenized, reduced in size, packed, labeled, and protected against contamination and decomposition. Insight into the foregoing theory and common sense will help in improving the quality of sampling. Although sampling hardly can be learned without actual practice, many books and papers on sampling practice are available (CH 78, CO 81, FO 85, KR 84, MI 89, SO 76). Directions for sampling specific products can also be found in the literature (AS 73, KE 88, ST 87a, WI 88).

2.7.1. Homogenizing the Sample

Often the sample is composed of increments. Because these increments by definition are not representative of the object as a whole, the sample is inhomogeneous. If the sample is to be analyzed as a whole this inhomogeneity is not important to the sampler, but usually only a part of the sample is used for analysis. Other parts are retained for duplicating the analysis, for countersamples, or for legal purposes. This implies that the sample will be sampled again, each sample subdivided into subsamples. Because the sample itself can be of the minimum size required, it is often not possible to take a subsample without further action. As follows from the theory of sampling inhomogeneous objects, the sample size is dictated by the number of composing parts or the autocorrelation of the object. There are two methods of obtaining a good sample that can be subdivided: (1) decreasing the autocorrelation, since a low value of the correlation constant permits small samples; and (2) decreasing the particle size, for instance, by crushing or grinding. Here as a rule homogenization is accomplished automatically. Homogenizing is also possible during the sample size reduction.

2.7.1.1. Small Samples

Small gaseous samples do not need homogenizing because usually the sample container is filled with a turbulent gas stream and this is sufficient to mix the contents. Furthermore, diffusion in gaseous samples is quite effective at room temperature, so storing the sample for some time suffices in situations where the homogeneity is questioned. If the sample container has been filled with a laminar gas stream or if for another reason the sample seems liable to inhomogeneity, it is possible to use mixing aids such as glass beads, metal beads, or small sheets of metal. When the sample container is revolved, these aids mix the gas quite easily.

Small liquid samples do not pose any difficulties whatsoever. Partly filled containers can be shaken by hand, and totally filled containers can be mixed on a mechanical roller. Sludges and fluids with suspended particles are by nature very difficult to homogenize. Here special instruments are mandatory. These instruments are like those used for subdividing granular material.

Small solid samples pose the most difficulties. Powdered or fine granular material can be mixed similarly to fluids, provided that the granules show a narrow particle size distribution. When fines are present, however, these tend to settle or agglomerate. The sample should be crushed or ground until a uniform particle size has been obained. This method is inapplicable when the particle size distribution of the object is the aim of investigation. Coarse samples of brittle material can be ground as well. It is recommended that a sieve be used during grinding, to separate the unground material from the already ground particles. The coarse fraction is recirculated until all material passes through the sieve. The size of the sieve is set by the rules for obtaining a good subsample and by the properties of the particles as far as segregation is concerned. This method saves time and, more importantly, does not bring the sample into contact with the material of the mill or mortar more than necessary. Besides mortars of china, agate, silicon carbide, tungsten carbide, or diamond, hammer or crushing mills are recommended. Ball mills should be used only when the analysis requires very fine samples.

Samples of metals are often in the form of turnings or shavings. Here homogenizing is very difficult because crushing or grinding is often impossible, except for the more brittle metals. The method of sampling should be modified to enable mixing of the sample before subsampling.

Materials that resist grinding or crushing, such as fibers or plastic materials, must be treated another way. Subsampling a sample of fibers can be very cumbersome. Unless specialized instruments, such as combs that make a fleece, are at hand, homogenizing must be done by "counting out." Every fiber or small bundle of fibers is assigned to a subsample, taking each subsample in turn and repeating this action until the whole sample has been treated and all fibers are divided between the subsamples.

Plastic material, like many polymers and foods, may often be kneaded on kneading rollers. However, when mechanical properties or chemical properties that are affected by kneading are involved, homogenizing is not possible and subsamples should be obtained in the form of multiple samples.

When the sample has to be used for description in detail, subsampling can pose problems. With solids it is difficult to state whether the fine structure is homogeneous throughout the sample. Here the best way is to estimate the correlation constant of the phenomenon sought and use a chunk of the material of a size that fulfills the requirements of a sample. Of course this can hardly be called homogenizing. In fact it is selecting subsamples in such a way that each subsample resembles a sample as far as possible.

2.7.1.2. Larger Samples

When sampling coarse or very inhomogeneous objects, the sample size can be considered if the sample has been taken according to the rules of Section 2.6. A sample of coke with a mean particle weight of 1 kg can easily be several tons. When air has been sampled, for instance, in assessing pollution, or when large lots of fluids or slurries are sampled, sample sizes of thousands of liters or kilograms are not uncommon. Of course the same rules as applied to small samples must be used here, but the execution is often quite different.

When voluminous gas samples must be collected, there are two possibilities. Samples under atmospheric pressure are often collected in inflatable bags. The content of a bag is very homogeneous because temperature differences, handling, and diffusion tend to cancel any inhomogeneity. When the sample is collected under pressure, whether this is done by compressing the sample or by sampling pressurized gas, homogenizing is difficult. The best way is to put some mixing aids in the pressurized vessel and homogenize the contents of the pressurized vessel by rolling. (Another way is to carry the container some hours in a car with a bad suspension on an equally bad road. Commuter vehicles are often more valuable for samples than for people.) As a rule, however, when large volumes of gas must be sampled, the sample is treated on the sampling site to collect the substance to be determined by extraction: drying filters, absorbing fluids, mechanical filters, or cold traps can be used. The contents of these aids can be treated like samples of fluid or particulate character.

Homogenizing large quantities of fluids is usually not difficult if the fluid is monophasic. If more phases are present, for example, two or more layers, emulsions, suspensions, or slurries, homogenizing can be a hard task. Usually efficient mixing and collecting of the sample while mixing suffice. However, for fluids with phases that differ greatly in density, "centrifugal" effects are a nuisance; in this case shaking is more effective than stirring.

Fluids that must be kept separated from the atmosphere—for example, condensed gases, oxidizable fluids, fluids liable to decompose with water, or fluids that have to be kept sterile—are always kept in closed vessels. If there is a headspace, and for safety reasons there should be, shaking suffices.

Homogenizing large quantities of solids is a difficult task. Usually this is accomplished by reducing the sample size and then grinding the entire sample.

For medium-sized samples (tens to hundreds of kilograms), various mixing devices that can shake the container in many complicated patterns can be useful, though this is no solution for samples of wide particle size distribution. A good method, though liable to bad results if not executed properly, is to collect chunks by hand (e.g., coal, ore, minerals, food) in such a way that each particle size is represented in the right proportion. This method can be excellent if those who collect the subsamples are aware of what they are doing: subsampling a sample in such a way that each subsample is like the gross sample.

2.7.2. Reduction of the Sample

As stated in the foregoing section, subsampling as a means to reduce the size of the gross sample to more manageable sizes or to obtain more samples for different purposes must be preceded by homogenizing.

Subsampling gases and liquids, when properly mixed, does not pose many difficulties. Subsampling solids, large samples, or samples with wide particle size distribution as a rule requires more sophisticated methods.

An old and widely known method is quartering. The sample is put in a heap, roughly mixed, and subdivided into four quarters. Two opposite quarters are eliminated, the remaining two are mixed again, quartered, and so forth, until the size of the remaining sample is manageable by other methods. However, the method of quartering is often used into the gram region.

A handy instrument for subdividing is the riffle, a slanted plane that contains separated partitions [Figure 2.20 (WO 79)]. By collecting only half of the alleys, the sample size is halved. Here the sample is not homogenized. Many subsamples with increment size $G = \frac{1}{2}A$ are collected and yield a gross sample. The procedure can be repeated by recirculating the collected (half) sample. During the "reriffling" a certain amount of mixing occurs. One instrument reduces the sample to about one-eighth to one-sixteenth of its original volume (three to four cycles). Loading the riffle with a part of the sample, eliminating the odd

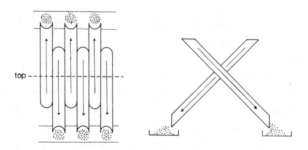

Figure 2.20. Riffle sampler.

loads, allows extended reduction. Use of a smaller riffle allows further reduction in the size of the sample. When the sample has been reduced in size sufficiently and is mixed by riffling, the riffle can be used to collect a number of subsamples.

Another mechanical means for sample reduction is the carousel. From a large hopper the sample is fed onto a string of smaller containers mounted on the circumference of a disk. Each container in turn receives a small part of the sample. The disk is revolved as long as sample flows from the hopper. The number of subsample containers determines the sample size reduction. Of course, other realizations of the same principle are possible, for example, a sample hopper revolving over the subsample containers. The carousel method can be used only for samples with a maximum size of a few kilograms. For larger samples the method has been incorporated into the sampling machine: the sample is fed onto a revolving disk. After each revolution a small section is swept into a sample collector; the remaining part is swept back into the product stream.

Note that from the preceding theory it follows that this type of sample reduction apparatus can be used only when some knowledge is available on the internal correlation of the samples. It is customary to homogenize the sample before reduction in order to make $p = P/T_x$ larger by reducing T_x (Section 2.6.1).

2.7.3. Sample Storage and Transport

After the sample has been taken or reduced in size, it has to be stored in a suitable container. The type of container used depends on how the sample is to be handled. In general it can be said that the container must be sturdy to protect the sample from mechanical damage during transport to the laboratory; the distance may be short but is often quite long. The sample may be kept for a fairly long time, and furthermore it should be protected from contamination by and losses to the environment. When the water content is to be determined, all water should be retained in the sample. A special case is the loss of trace components. Organic components that are present in minute quantities, for example, in natural waters, can be lost by adsorption onto the surface of the container, by adsorption into rubber stoppers or sealing rings, by oxidation or decomposition under the influence of light, or by evaporation and diffusion through container walls.

Trace metals can be lost by adsorption on the wall of the container, and mercury can be lost by evaporation of the metal or, more likely, its organic complexes. A rather special cases is the loss of sample components by biological processes: rodents and insects can affect samples of food, and bacteria and fungi may alter not only food but also many organic components and even inorganic salts.

Another cause of sample alteration can be human interference. In particular, when samples are used for trade or legal purposes it is important that no changes whatsoever can be caused by humans.

This leaves only a few types of suitable containers. When dry solids are to be packed and their water content is not important, paper bags, plastic bags, or a combination of these are used. For trade or legal samples provision should be made for the container to be sealable. A good overview of sampling equipment is given by Keith (KE 88) and Sturgeon (ST 87b).

Glass is one of the most suitable materials, either as glass jars with a screw cap or flasks. Most solids can be kept in glass containers for a sufficient length of time. Air and water can be excluded; fines and water are retained. Most solids (except some strong alkalies) do not affect glass.

Most liquids can be kept in glass also. As already mentioned, special care is required when the sample contains trace components that should be determined. Since the adsorption of metals is probably caused by ion exchange effects, a low pH of the sample seems to be a prerequisite. Care must be taken that no components from the glass will contaminate the sample. In general, traces of sodium, silicon, and calcium are dissolved in water when stored in glass.

Gaseous samples may also be kept in glass containers. However, since the container should now be provided with stopcocks, sealing and transport are difficult. Here metal containers are often used. When compressed gases that should be kept under pressure and sampled, metal is obligatory.

Plastic containers, especially poly(ethylene) flasks and jars, are used extensively for transporting samples. The advantages are obvious—they are unbreakable and light—but the disadvantages are the adsorption of trace organics, often irreversible, and the adsorption of trace metals.

Contamination of samples with the plastic, plasticizer, or traces of catalyst is often encountered, especially when organic liquids are stored in plastic bottles.

Gases can be kept in plastic containers only for short periods, since many gases can permeate through thin plastic walls to both sides, so loss of components as well as contamination with oxygen or water are possible.

Sealing of containers against loss or entry of components by using screw caps with seals, rubber stoppers, or Teflon-lined stoppers is usually sufficient. Glass stoppers and cork stoppers should be avoided except when used for dry, coarse solids. Sealing of containers for trade or legal purposes is best done by locking a screw cap with a thread and sealing the knot.

For special purposes, for example, packing samples of highly corrosive or toxic substances and radioactive materials, the appropriate directions should be followed, as given in the literature.

Many samples of food products and samples from the environment are liable to decomposition by bacteria, molds, or enzymes. The most universally applicable treatment is cooling. Samples can be packed in solid carbon dioxide or stored in a freezer. Sterilizing by heat is seldom used. Preservation by adding a bactericide (thymol, toluene, or mercuric chloride) is often used for environmental samples that are taken in the field or at unattended sampling stations.

A sample is valuable only when it represents the lot it is taken from. This implies that the lot from which it originates must always be known. Because the sample itself as a rule cannot bear the label with its origin (exceptions are samples such as ingots or packed items of food or fine chemicals), the container should bear this information. All data can be recorded on a label firmly attached to the container, or the data can be held in a protocol coupled to the container by an identification number.

The sample is identified unambiguously if the time and place are accurately registered. In practice other information may be provided, such as an identification of the lot (e.g., the name of a patient, the name of a ship containing the lot, or the statement of a traded lot). It is important that the sampler, whether human or machine, can always be identified. In routine work this can be traced if the time and place of sampling are known. In all other situations the sampler should sign the label or the protocol.

Samples for trade or legal purposes must be sealed by the sampler, who should sign the protocol with a witness. In order to circumvent problems when subsamples are being made for storage or contra-expertise, at least three samples must be provided, separately packed and labeled. Here the sample container often may contain only an identification number, but more data should be recorded in the protocol.

One of the most serious mistakes that can (and will) be made in sampling and analysis is the inadvertent interchange of sample and data. The only infallible method for avoiding such mishaps is to distinguish samples by an inherent property, for example, composition. Many ingenious systems have been developed to prevent mistakes, especially in clinical chemical laboratories. Most systems rely on the identification at the sampling point being correct and stress retention of the identification mark with the sample, for example, by mechanical reading and duplication of labels and samples or by preventing duplication errors by using error-detecting identification numbers or parity bits.

2.8. QUALITY CONTROL OF SAMPLING

As noted in the preceding sections, sample quality can be affected in many ways. The most important sources of sample deterioration are inappropriate sampling schemes (sampling frequency, increment size, or number of increments), inappropriate sample handling (loss of components, contamination, or decomposition), and inappropriate sample identification (mislabeling and erroneous coupling of sublabels to subsamples).

Forward control can be achieved by proper training of samplers and a thorough setup of sampling schemes. Feedback control is possible by analyzing duplicate samples that are obtained routinely or by random instructions.

Analysis of variance (Section 4.6) may indicate whether the error found is caused by sampling or by analysis.

CHAPTER

3

ANALYSIS

3.1. CHARACTERISTICS OF AN ANALYSIS

A chemical analysis of a material gives a characterization of that material, or a sample of that material, in terms of chemical composition. A distinction may be made between qualitative and quantitative chemical analysis. A qualitative chemical analysis describes the sample in terms of the identity of the composing elemental parts—atoms or molecules—whereas a quantitative chemical analysis also gives the quantities of each of the composing parts of the sample. Another distinction between analytical analyses can be made based on the type of the characterization: the sample can be described according to the occurrence of types of elements—sorts of atoms—which can be qualitative as well as quantitative. This characterization can be extended with information providing the structure of the sample. Here the link between the various types of atoms is involved, and this type of analysis can be considered as a special kind of qualitative analysis. Apart from chemical analysis, physical analysis is also applied for description of samples. Sometimes it is hard to distinguish between chemical and physical sample analysis because a chemical composition is often correlated to a physical quantity, for example, in the typification of polymers and other macromolecular compounds. On the other hand, sometimes a chemical analysis is made in order to give a physical description of a sample.

3.1.1. Structure of an Analytical Procedure

The analytical procedure nearly always consists of six distinguishable steps:

1. Sampling
2. Sample preparation

3. Measuring
4. Data processing
5. Testing, controlling, and eventually correcting one or more of the processing stages
6. Establishing a quality merit

The sampling is described in Chapter 2 together with the considerations involved in a proper sample preparation; the measuring process and the various ways data may be processed are discussed in Chapter 4. The only thing to be mentioned here is that no analytical procedure is complete without a proper validation of each of the stages and of the final result. Based on the type of measurements, analytical processes can be divided into one-dimensional and two-dimensional methods. In spectral analyses where qualitative information on the sample is available beforehand, sometimes the measurement of a single spectral position is sufficient to provide a concentration as well as the complete spectrum. In practice many quantitative analyses are one-dimensional methods. It is seen that each individual compound or element in a sample requires at least a one-dimensional analysis. Sometimes information on quantitative properties may yield an identification of a sample by a single-value method. For instance, comparison of the measured refractive index of a very pure chemical compound with a file that correlates refractive indexes with molecular structures may provide identification. As a rule the reliability of such an identification is not very high and more one-dimensional methods for other typification should be applied.

Another possibility is a series of measurements as a function of a parameter that characterizes the method. In the continuous wave measuring of a spectrum the energy scale (wavelength, wavenumber, or magnetic field) is varied in a well-defined and selected manner and at each energy value a measurement (e.g., absorbance or magnetization) is made. The resulting set of data pairs can be represented in a graphic way (Figure 3.1), with y (measured quality) being a function of x (selected energy value). Other representations such as dy/dx versus x may also be used, but this does not change the principle of the two-dimensional method. In general, qualitative information is found from the x-axis and quantitative information is found from the y-axis of the graph.

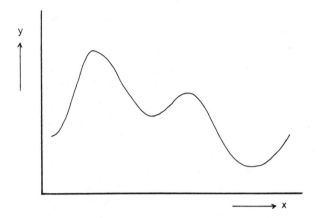

Figure 3.1. Measured quality, y, as a function of energy value, x.

In two-dimensional methods two different physical or chemical parameters are scanned. At the exit of an HPLC (high-performance liquid chromatography) instrument a diode-array detector may be mounted. This results in a full spectrum at each point of the chromatogram, of course with a limited, though high, resolution. Now the result of the analysis is a matrix. The information content is very high, in fact too high for full use. The excess information can be used however. Increasing the precision is one way (more data gives more possibilities to diminish noise). Another way is to use the extra information to diminish the physical cost: less separation of the peaks in the chromatogram can be compensated by calculation using the spectral information (Section 4.8). Other examples are GC–MS (gas chromatography–mass spectrometry; more qualitative information from chromatographic peaks), MS–MS (better separation possibilities), etc.

3.2. QUALITY OF AN ANALYTICAL PROCEDURE

In the foregoing section some characteristics of and various distinctions between analytical procedures are made. Given an analytical problem, before a decision can be made on how to select the most appropriate method, more method features must be identified. This discussion aims for a detailed description of the quality of an analytical procedure. A set of features describing quality aspects of analytical procedures are given and defined in the literature (KA 73, PO 92, VA 77a). Some of these properties of features cannot be discussed without also discussing the chemical problem to which the procedure is applied. A method providing a very low detection limit and high accuracy and reproducibility may be adversely influenced because of low selectivity. In practice this may not hamper its application provided that the sample preparation step preceding the measuring step removes all disturbing elements. Thus the effect of poor selectivity is to require selective sample preparation or application to problems in which other constituents of the sample matrix play no role. Specific methods with high accuracy and reproducibility may require sample quantities that are relatively high. Here, too, a preprocessing step on sample preparation can be of help, for example, by preconcentrating a sample.

This technique can be applied successfully in some types of chromatography where a preconcentration column is used to sample the analyte from a large amount of sample of low concentration. The analyte thus obtained is backflushed under other physical or chemical conditions that are applied during the sampling, such as higher column temperatures or elution with a gas or liquid stream that has a higher affinity toward the stationary phase than the analyte.

Problems arising from low selectivity or specificity are met when the response of the measuring device is caused by the analyte as well as by one or more components of the sample matrix. In situations where the sample preparation cannot give a proper separation between the analyte and the disturbing components an extended calibration procedure combined with a calculating technique may solve the problem, provided that the number of disturbing components is not too high and the signals of the measuring device are well above the noise level (HO 78, LO 84, MA 78a, MA 88a). With reference to the cost of an analysis it should be mentioned that, based on salaries of laboratory personnel and prices of modern analytical instruments, a sample preparation of 15–20 minutes by laboratory technicians usually costs considerably more than a measurement of the same period using an advanced instrument (see Table 3.9 in Section 3.3.1).

These circumstances and the need for quick and reproducible sample analysis, such as is imperative for real-time control of chemical processes with a low time constant, may demand automatic analytical

Table 3.1. Completeness of Published Analytical Methods

Positive Evaluation	Paper (%)	Abstract (%)
Condition analyte	50	40
Analytical range	50	20
Detection limit	15	5
Optimization	60	—
Purity + brand	15	—
Reagent storage	20	—
Sample preparation	75	30
Main reagent	90	65
Blanks	50	—
Precision	30	5
Accuracy	10	5

procedures even at high initial instrument costs. In research laboratories, where time and personnel situations are considerably less important than in industrial plants, another analytical procedure may be preferred or selected.

Laboratories that use routine methods have their time-proven or standardized methods of analysis. However, often new methods are required, for instance, in research laboratories or in routine laboratories in a changing environment. In that case new methods are sought in literature. Unfortunately neither abstracts nor full articles provide enough information. A thorough check and more information are always required. Postma et al. (PO 92) analyzed a number of descriptions of analytical methods and found the results shown in Table 3.1.

All the remarks of this section should be kept in mind when selecting an analytical procedure, discussed later in Section 3.3.

3.2.1. Limit of Detection

Whenever a sample containing a compound in a very low concentration has to be measured by an analytical procedure, the signal from the measuring instrument as a rule will be small. It is difficult to decide whether the signal emerges from the component to be determined or from the inevitable noise produced by the procedure or the instrument.

This uncertainty gives rise to the so-called limit of detection. The quality of this limit can be determined by statistical means.

When one is deciding whether a measured signal originates from the measured property or does not, some incorrect decisions can be made:

1. A decision that the component was present in the sample when in fact it was not: error of the first kind
2. A decision that the component was not present in the sample when in fact it was: error of the second kind

Furthermore, it can be argued that there will be a signal that is an indication of the presence of the component, but that does not allow us to make a statement about the true amount of the component. The incorrect decision can now be the following:

1. A decision that the amount of component can be given with some precision, when this is not true
2. The opposite decision—that the amount of component cannot be established when in fact this is possible

In defining the limit of detection a criterion must be selected that is applicable to the decision whether a signal can be used for stating that a component is present or not present and, if present, with what reliability it can be quantified. There are many definitions given in the literature. Some, however, using the same criterion, claim different confidence levels. Others define the same limit of detection but use different criteria. As a rule the definition of the limit of detection is based on errors of the first kind.

The limit of detection is the smallest observed signal (x) that with a reliability $1 - \alpha$ can be considered as being a signal caused by the component to be measured. When the observed signal is smaller than x, however, it cannot be stated that the component is absent. It can only be said with a reliability $1 - \beta$ that the concentration of the component will be less than a certain value c_G.

If a blank (concentration of the component to be known, c_0) and a sample (concentration c_1) are analyzed repeatedly, the measuring signals obtained are distributed around $\bar{x}_0$ (blank) or $\bar{x}_1$ (sample) (Figure 3.2).

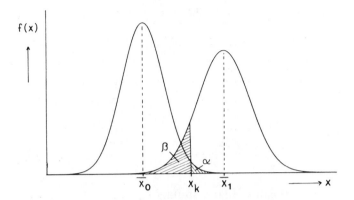

Figure 3.2. Limit of detection: $\bar{x}_0$ = mean signal of blank; $\bar{x}_1$ = mean signal of sample; x_k = criterion value; α = probability that signal of sample is considered as signal of the sample; β = probability that signal of sample is considered a signal of blank.

If the signal with magnitude x_k is used as a criterion for the presence of component c, the probability α that an observed signal $x > x_k$ caused by the blank is

$$\alpha = \int_{x_k}^{\infty} P_0(x)\,dx \qquad (3.1)$$

where $P_0(x)$ = probability distribution of x_0
     $P_1(x)$ = probability distribution of x_1

The probability β that a signal will be observed that is less than x_k and caused by the sample (concentration component x_1) is

$$\beta = \int_{0}^{x_k} P_1(x)\,dx \qquad (3.2)$$

where α = probability of an error of the first kind
     β = probability of an error of the second kind

If $P_0(x) = P_1(x)$ and $\alpha = \beta$, then x_k is the point of intersection of the two probability distributions.

Now the limit of detection x can be defined as

$$x = \bar{x}_0 + k\sigma_0 \qquad (3.3)$$

Table 3.2. Value of k

α	k
0.5000	0
0.1587	1.0
0.0228	2.0
0.0062	2.5
0.0026	2.8
0.0013	3.0
0.0005	3.3

where $\bar{x}_0 = $ mean of blank signals x_0

$\sigma_0 = $ standard deviation of the probability distribution $P_0(x)$ of x_0

$k = $ factor fixed by $P_0(x)$ and the preferred value of $1 - \alpha$

The composition derived from the signal x is

$$c = \frac{\bar{x}_0 + k\sigma_0}{S} \tag{3.4}$$

where $S = $ sensitivity (Section 3.2.2).

The value of k can be obtained in various ways:

1. When the the probability distribution x_0 is known with sufficient confidence and proved to be Gaussian, α and k are related as shown in Table 3.2. An often-used value of $\alpha = 0.0013$ $(1 - \alpha = 99.87\%)$ results in $k = 3$, so that

$$\mathbf{x} = \bar{x}_0 + 3\sigma_0 \tag{3.5}$$

2. When σ is not known but estimated a s from a finite number of observations, Student's-t must be used (t is a correction factor used to compensate for the uncertainty in the estimation of s) (Section 3.2.7).

When for a certain method of analysis the limit of detection x is set equal to the mean of the blank signal,

$$\mathbf{x} = \bar{x}_0 \tag{3.6}$$

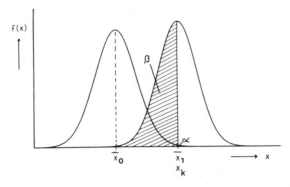

Figure 3.3. Limit of detection. See legend to Figure 3.2 for symbols.

then $k = 0$ and $\alpha = 0.5$. This implies that a random result is obtained. Now the probability that a signal of a concentration equal to the blank will be considered as a signal of the component is 50%. If the obtained signal $x = \mathbf{x}$ and $k = 3$, we know that

$$x = \mathbf{x} = x_0 + 3\sigma \qquad (3.7)$$

The probability α that a signal caused by the blank will be considered as a signal caused by the component is 0.0013 (Figure 3.3) (assuming $\sigma_0 = \sigma_1$); β, the probability that a signal caused by the component will be considered to be caused by the blank, is now 50%. If the mean observed signal $x < \mathbf{x}$, we cannot state that $c = Sx$ is smaller than C, the

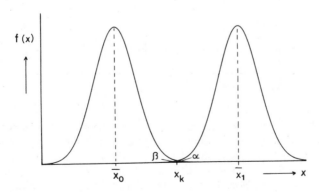

Figure 3.4. Limit of detection. See legend to Figure 3.2 for symbols.

concentration in the blank, because the reliability of this statement is only 0.5. If a signal $x_G = \bar{x}_0 + 6\sigma_0$ (assuming $\sigma_0 = \sigma_1$) and $x = \bar{x}_0 + 3\sigma_0$, then the situation shown in Figure 3.4 exists.

For a signal $x = x$ it can be said that the concentration in the sample is less than c_G (if $c_G = Sx_G$) with a probability of 99.87%. This value $c_G = Sx_G$ is sometimes called the "limit of guarantee for purity." Here c_G is the lowest concentration that can be detected with a probability of 99.87%.

If $\sigma_0 \neq \sigma_1$, x_G is given by

$$x_G = \bar{x}_0 + 3\sigma_0 + 3\sigma_{xG} = x + 3\sigma_{xG} \tag{3.8}$$

Sometimes a limit of determination is established as

$$x_D = \bar{x}_0 + 10\sigma_x \tag{3.9}$$

Currie (CU 68) distinguishes three areas (Figure 3.5):

  I. The area where detection is not possible at all (tossing a coin would give the same information) or possible only with a limited probability ($\bar{x}_0 < x < x_G$)

  II. The area where detection is possible with a sufficient probability ($x > x_G$)

  III. The area where quantitative determination is possible ($x > x_D$)

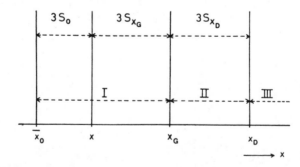

Figure 3.5. Limit of detection (CU 68): (I) the area where detection is not possible; (II) the area where detection is possible with a sufficient probability; (III) the area where quantitative determination is possible. Here $\bar{x}_0 =$ mean of blank signals x_0; $x =$ limit of detection; $x_G =$ limit of guarantee of purity; $x_D =$ limit of determination. Reprinted with permission from *Anal. Chem.* **40**, 588 (1968). Copyright by the American Chemical Society.

Often the blank and the sample are analyzed simultaneously. The signal of the blank is used to correct the signal of the sample and also to estimate $\bar{x}_0$:

$$x_S = x_{S+x_0} - x_0 \tag{3.10}$$

and

$$\sigma_{xs} = (\sigma_{xs+x_0}^2 + \sigma_0^2)^{1/2} \tag{3.11}$$

If $\sigma_{xs+x_0} = \sigma_0$ then $\sigma_{xs} = \sigma_0\sqrt{2}$. Now $x = \bar{x}_0 + k\sigma_0\sqrt{2}$. If more determinations are made (n), then

$$x = \bar{x}_0 + \frac{k\sigma_0\sqrt{2}}{\sqrt{n}} \tag{3.12}$$

This procedure is applied for enhancing signal-to-noise ratios at the expense of analysis time. In the literature various names are used for the criteria discussed here, as shown in the accompanying tabulation.

x	Currie:	a posteriori upper limit ($k = 1.64$)	(CU 68)
	Kaiser:	detection limit ($k = 3$)	(KA 70)
x_G	Currie:	a priori detection limit ($k = 3.29$)	
	Kaiser:	limit of guarantee for purity ($k = 6$)	
	Liteanu:	detection limit	(LI 73, LI 75)
	Svoboda:	*Nachweisgrenze*	(SV 68)
x_D	Currie:	determination limit ($k = 10.1$)	
	Kaiser:	determination limit ($k = 10$)	

In the foregoing the values of α and β have been established on the assumption of a Gaussian distribution of the measurements. In practice values of α for $k = 3$ will not be 0.0013 as calculated but much higher, even up to 0.05. This is because the distribution of most determinations in this low concentration range is non-Gaussian. Furthermore, the inhomogeneity of the sample can be quite large in this region, especially when the distribution of the component to be analyzed is not homogeneous but particulate.

A survey of the methods of estimation of the limit detection has been given by Ingle (IN 74b), Long and Winefordner (LO 83), and Boumans (BO 78). Hirschfeld (HI 76) gives an account of the practical aspects of limit of detection. For skewed data a transformation can be obligatory (Section 4.1.2). In X-ray spectrometry and other analytical methods based on particle counting, the limit of detection depends more on the counting time than on the background and blank values.

3.2.2.   Sensitivity

A procedure such as an analytical method with one input (the sample) and one or two data outputs can be characterized by the sensitivity S, which is the ratio of the quantitative output and the input, for a given qualitative range. For example,

$$S(\lambda = 534\,nm) = y/x \tag{3.13}$$

where $y =$ extinction at 534 nm
$\qquad x =$ concentration of component x

Often, however, y is composed of a part that depends on x and a part independent of x (blank). Furthermore, as a rule y is not linearly proportional to x over the entire range of possible x and y values. Therefore there are good reasons to define

$$S(x, y) = dy/dx \tag{3.14}$$

for the range of x and y values where this relation holds. The range for which S exists and has an unambiguous value is the *dynamic range* of the procedure. Mostly for reasons of convenience, analytical chemists try to develop methods in which S has a constant value in a range as large as possible. This range is called the *linear dynamic range* and is expressed in the orders of magnitude for which s can be considered to be a constant.

The dynamic range is limited at the lower level by the value of x where y cannot be distinguished from the noise in y. Here S may have all possible values. The upper limit of the dynamic range will be set by saturation of the detector, $S \to 0$. The linear dynamic range is often defined as the region where $|y - Sx'| \leqslant 0.03Sx'$, y' being the true signal and Sx' the computed signal, assuming a linear relation between y and x (Figure 3.6).

In situations where the analytical procedure consists of a chain of procedures, the sensitivity S can be subdivided into sensitivities belong-

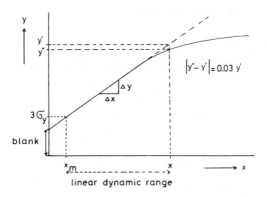

Figure 3.6. Sensitivity and linear dynamic range for output signal y as a function of input signal x.

ing to the consecutive operations (Figure 3.7):

$$S_{\text{total}} = \frac{dy_n}{dx_n} \cdot \frac{dx_n}{dx_{n-1}} \dots \frac{dx_2}{dx_1}$$

$$= \frac{dy_n}{dx_n} \cdot \frac{dy_n}{dx_{n-1}} \dots \frac{dy_1}{dx_1} \tag{3.15}$$

$$= S_n S_{n-1} \dots S_1$$

$$= \prod_{1=1}^{n} S_i$$

Note that the dimension of S depends on the dimension of x and y. If, for example, x is given in grams and y in milliamperes, the sensitivity S is expressed in mA/g.

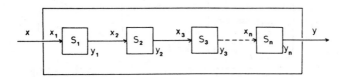

Figure 3.7. Cumulative sensitivity for a signal y resulting from a number of consecutive operations S_i on an input signal x.

Consider the chain of procedures:

- Concentration (g/L) converted into a potential (mV): $S = mV/(g/L)$
- Potential converted into deflection of a recorder pen (cm): $S = cm/mV$

Then the ultimate sensitivity is expressed in

$$S = \frac{mV}{g} \cdot L \cdot \frac{cm}{cV} = \frac{cm \cdot L}{g} \tag{3.16}$$

The foregoing holds only for static measurements. Under dynamic circumstances S is time dependent. A sudden disturbance of the value of x is not momentarily followed by an output y; dy/dx is a function of t.

If we assume that y is related to x according to a first-order differential equation, a stepwise disturbane of $0 \rightarrow x$ results in

$$y(t) = Sx[1 - \exp(T_x/\tau)] \tag{3.17}$$

where T_x = time constant of the procedure. This results in a response like that shown in Figure 3.8.

Because the output y is time dependent, there will be a bias between

$$y(t \rightarrow \infty) = Sx \tag{3.18}$$

and

$$y(t) = Sx[1 - \exp(-T_x/\tau)] \tag{3.19}$$

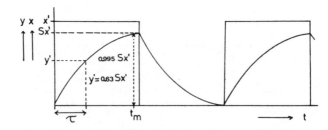

Figure 3.8. Influence of time constant τ on response y of repetitive analysis.

If this bias should be limited to a fraction Δ of the maximum signal, then

$$y(t \to \infty) = (1 - \Delta)y \tag{3.20}$$

According to equations 3.18 and 3.19

$$(1 - \Delta)Sx = Sx[1 - \exp(-T_x/\tau)] \tag{3.21}$$

and

$$T_x = -\tau \ln \Delta \tag{3.22}$$

For $\Delta = 0.01$, T_x must be at least 4.6τ. However, the measuring instrument has to return to zero for $x \to 0$, which means that the time lag between two subsequent readings must be at least $T_x = -2\tau \ln \Delta$. This condition implies that for $\Delta = 0.01$, $T_x \approx 10\tau$. Many instruments of the type that convert a physical quantity into a current or a potential have time constants that range from 100 to 1000 ms. Instruments incorporating handling of mass, such as chromatographs and flow analyzers, have time constants ranging from 100 ms to 10 s. Here the time between subsequent analyses can be on the order of 100 s. The foregoing remarks apply only to instruments having a first-order response. If time delays are incorporated, as is usual during manipulation of the sample, the time lag between two successive analyses is the sum of this delay time plus the response time of the meter.

3.2.3. Safety

The aspect of safety in analytical chemical practice is very difficult to incorporate quantitatively into the set of criteria. Analytical instruments as a rule are safe provided that they satisfy the safety requirements for electrical instruments. High-energy spectroscopes, emitting X-rays, are liable to give off dangerous radiation, as do instruments fitted with radiation sources, such as gas chromatographs fitted with γ-ray detectors based on Li^3H. If they are properly handled and the rules set for radioactive materials are satisfied, these instruments do not have a safety risk that need be considered in deciding whether to apply an analytical method or not.

The procedures as a rule are safe. Possible exceptions are the destruction of organic material by fusion in sealed vessels or by perchloric acid. If, however, the procedure is carried out according to the directions of the safety standards and the required safety precautions are applied, no abnormal risk is to be expected.

The only risk factor that seems to be important when one is choosing a method of analysis is the use of dangerous reagents in the preprocessing of the sample. These can be radioactive (in isotope dilution or in the processing of irradiated samples from activation analysis), extremely poisonous, or carcinogenic reagents. Such risks can be eliminated to a large extent either by the use of special equipment such as hoods and glove boxes in special procedures like decontamination or by specialized training required by "good laboratory practice," but in most cases the cost will be high. When one is selecting a method it seems appropriate to translate the safety risk into cost, which is determined by the measures required to reduce the risk to an acceptable level.

3.2.4. Cost

Cost as a quality criterion seems a bit odd, but if the criterion used is a "desired property," cost can be a discriminating factor. Cost can be subdivided into several parts and in various ways: overhead and allocated, fixed and variable.

Overhead costs are those of the setup required for analysis, comprising the laboratory building, joint instruments, service facilities such as the instrument repair shop and storage, utilities, and nonallocated personnel such as the director, secretary, and cleaning people. Often the overhead costs are not allocated to the laboratory but are fixed costs. Of course somebody has to pay these costs, but as a criterion of the quality of a method of analysis they are not appropriate. Allocated costs are those costs that can and will be paid by the customers of the laboratory. Whether or not these costs are really paid by the customer, they can be used as a criterion to describe the quality of a method of analysis.

Fixed allocated costs are those that are primarily caused by the method of analysis, not directly by the number of analyses. Often a trade-off is possible between fixed and variable costs. Mechanization and automation may increase the price of an instrument considerably, but as a rule the manpower needed decreases with increasing degree of automation. The optimum depends on the price of labor, including the unmeasurable benefits of replacing dull work and its increased risk of mistakes. Furthermore, the number of analyses per period and the price of the instrument are important. Another important trade-off is possible in optimizing the cost of accuracy. A rule of thumb is the prices of instruments increase quadratically with a rise in accuracy. Because accuracy can be increased by more frequent analyses or by more cautious (as a rule, slower) analyses, here again the price of labor and instrument should be incorporated.

In analysis for control, the most important way of optimizing costs, and thus seeking the most appropriate variables, is to compare costs of analysis for control with the positive or negative costs of the manufacturing process as a result of control. In Section 2.6.4 the influence of accuracy and frequency of analysis on costs was treated.

Data on costs of analysis are scarce in the analytical literature. Prices of instruments differ very much in different countries, as do labor costs. Moreover, the time required to perform an analysis is seldom given. A rough estimation of the time required between introduction of the sample and release of the analytical result can be obtained by analysis of the method description.

Often the time required for "passive" manipulation such as boiling, waiting, or cooling is described, for instance, "keep boiling for 5 minutes," or "leave on water bath for 5 hours." However, these times are labor extensive. A rough estimation of the times required is possible but is of little use. "Active" manipulations in routine or semiroutine analysis are divided into three categories, the number of manipulations in each class being counted and multiplied with a standard time for each class. If some nonallocated time is spread over the manipulations, the values of Table 3.3 are often valid. Exceptions are analyses by NMR (nuclear magnetic resonance), MS, GC, HPLC, and IR (infrared), where the time of analysis depends strongly on the type of measurement.

It must be borne in mind, however, that costs of analysis are very much influenced by the organization of the work. Not only overhead but also "friction losses," waiting time when queues develop, and training and standard of training of personnel, for instance, make possible differences in costs so large that their reliable use as a quality criterion is very difficult.

When analysis time and waiting times are taken into consideration, cost can be estimated to a certain extent. Laboratory simulation

Table 3.3. Classification of Analyses[a]

Manipulation	Class	Standard Time (minutes)
Addition, transferring, mixing	1	2.5
Weighing, crushing, dissolving, filtration, heating, cooling, titration	2	5
Extraction, measurement with AAS, FES, UV-vis, AES, XRF	3	15

[a] AAS = atomic absorption spectroscopy; FES = field emission spectroscopy; UV-vis = ultraviolet–visible range (spectroscopy); AES = atomic emission spectroscopy; XRF = X-ray fluorescence (spectroscopy).

(Section 5.1.1.3) can help in the setup of a cost-effective organization by optimizing batch size, choosing between automated or nonautomated analysis, and setting priorities in a more or less objective way. Cost as a function of the number of samples taken to establish a fixed precision has already been considered. The concept of *critical batch size* is discussed as a measure useful in deciding whether a laboratory procedure should be automated or mechanized. Dependence of critical batch size on initial investment, cost of reagents, frequency of sample analysis, lifetime of the equipment, and other important factors should be considered.

Cost is also mentioned in Sections 2.6.3, 2.6.4, 3.1, 3.3.1, and 5.2.1.

3.2.5. Measurability

In studying processes one is interested in an estimation of the quality of equipment and procedures that are applied for measurement and for process control. Criteria that may allow an estimation of this quality are measurability and controllability. In Section 2.6.5 the controllability is given by the numerical reduction of the value of a disturbance of the process σ_x by a controlling action. The remaining value of the disturbance after control, σ, and that before define the controllability factor r according to

$$r^2 = \frac{\sigma_x^2 - \sigma^2}{\sigma_x^2} \qquad (0 \leqslant m \leqslant 1) \qquad (3.23)$$

In order to control a process, that is, minimize the difference between a measured quantity and its desired value $\bar{x}$, one must measure that quantity. The measurability m has been defined in a way resembling the definition equation of the controllability factor:

$$m^2 = \frac{\sigma_x^2 - \sigma_{min}^2}{\sigma_x^2} \qquad (0 \leqslant m \leqslant 1) \qquad (3.24)$$

in which σ_{min}^2 equals the variance of the measured value after application of ideal control. A nonideal control results in a σ value higher than σ_{min} and thus $r < m$, or, as stated in Section 2.6.4,

$$r = r_p m \qquad (r_p \leqslant 1) \qquad (3.25)$$

The aim of a controlling action is the reduction of disturbances of the process values. The disturbances can be characterized by means of statistical parameters such as

- A distribution-density function that characterizes the amplitudes of periodic process disturbances
- A correlation function that describes the frequencies of periodic process fluctuations

The standard deviation of the disturbance σ_x before control and σ after control are related according to

$$\sigma = \sigma_x (1 - r^2)^{1/2} \qquad (0 \leqslant r \leqslant 1) \qquad (3.26)$$

A value $r = 0.5$ results in $\sigma = 0.87\sigma_x$, which implies a very small reduction of the disturbances. A good controlling action produces an r value > 0.9, which means 80% suppression of disturbances.

The frequency of the process fluctuations follows from the autocorrelation function of x_t. According to Section 4.5.2, the autocorrelation gives a relationship between a process value of time t and that of another time. This predicting element can be used for measurability and controllability (MU 78a).

When the result of a measurement has been obtained, a prediction can be made of the value of x in the future, a period τ later.

The uncertainty in this prediction is given by the predicted variance

$$\hat{\sigma}_x^2 = \sigma_x^2 (1 - \alpha^{2\tau}) \qquad (3.27)$$

For first-order, stationary stochastic processes $\alpha = \exp(-1/T_x)$, $\hat{\sigma}_x^2$ equals the predicted variance, and σ_x^2 the variance known from the history of the process.

When the process is sampled regularly, with the time between two consecutive samples, T_a, and the time required to obtain the result of the analysis, T_d, it can be stated that the uncertainty is minimal at the moment the result is produced, a period T_d after the moment of sampling. The uncertainity gradually increases during the period T_a and sharply decreases when the next result is obtained.

The mean predicted variance $(\hat{\sigma}_x^2)_m$ over this period is

$$
\begin{aligned}
(\hat{\sigma}_x^2)_m &= \frac{\sigma_x^2}{T_a} \int_{T_a}^{(T_a + T_d)} \left[1 - \exp\left(\frac{-2t}{T_x}\right) \right] dt \\
&= \sigma_x^2 \left[1 - \exp\left(-\frac{2T_d}{T_x} \right) \left(\frac{1 - \exp(-2T_a/T_x)}{2T_a/T_x} \right) \right] \\
&= \sigma_x^2 \left[1 - \exp\left(-\frac{2T_d}{T_x} \right) \right] \exp\left(-\frac{T_a}{T_x} \right)
\end{aligned}
\tag{3.28}
$$

$$
m^2 = \frac{\sigma_x^2 - (\hat{\sigma}_x^2)_m}{\sigma_x^2} = \exp\left(-\frac{2T_d}{T_x} \right) \exp\left(-\frac{T_a}{T_x} \right)
\tag{3.29}
$$

$$
m = \exp\left(-\frac{T_d}{T_x} \right) \exp\left(-\frac{T_a}{2T_x} \right)
\tag{3.30}
$$

By the same resoning the expression for m has been derived (GR 65, GR 66) when the accuracy σ_a of the measuring method and the sample size T_g is included:

$$
m = \exp\left(-\frac{T_g}{3T_x} \right) \exp\left(-\frac{T_d}{T_x} \right) \exp\left(-\frac{T_a}{2T_x} \right) \left(1 - \frac{\sigma_a T_a^{1/2}}{\sigma_x T_x^{1/2}} \right)
\tag{3.31}
$$

or, for short,

$$
m = m_g \cdot m_d \cdot m_a \cdot m_n
\tag{3.32}
$$

Figure 3.9 gives the m values as a function of T_p/T_x and T_d/T_x, and Figure 3.10 gives the m values as a function of σ_a/σ_x and T_a/T_x. These

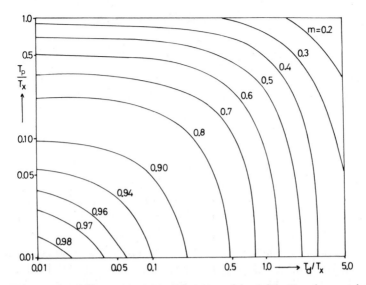

Figure 3.9. Measurability constant m as a function of dead time T_d and measuring time interval T_p for a first-order autoregressive stochastic stationary process with a time constant T_x.

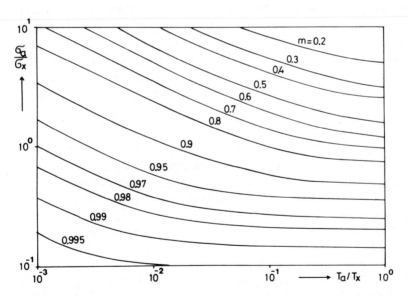

Figure 3.10. Measurability constants m as a function of accuracy of the measuring method, σ_a and time lag between two consecutive samples T_a for a first-order autoregressive stochastic stationary process with a time constant T_x and a standard deviation σ_x.

Table 3.4. Reduction of Disturbances

σ_{min}/σ_x	m
0.10	0.995
0.25	0.97
0.50	0.87
0.87	0.5

two figures imply a strategy for improvement of m by an attack on the smallest value of m_g, m_d, m_a, or m_n.

One must keep in mind that all equations presented here are approximations that are valid for high values of m only, but in situations encountered in practice this is hardly a disadvantage because low m values imply $(1 - m^2)^{1/2} \approx 1$, so the variance decrease is negligible and thus it is immaterial that m may have an approximated value. This is illustrated in Table 3.4, where the reduction of disturbances expressed as σ_{min}/σ_x is tabulated as a function of m according to eq. 3.27:

Ivaska (IV 86) estimated the optimal sampling frequency for process analyzers, starting from the Nyquist theorem: the sampling frequency should be at least twice the highest frequency of the studied signal. This sets limits to the lowest allowable sampling frequency of a process analyzer. Economic criteria together with the instrument analysis time define the highest sampling frequency. Introducing a number of constraints, the resulting rule of thumb is that the sampling time should be equal to the dominant time constant of the system.

Example. The practical implications of these equations can be illustrated with the following example pertaining to the analysis of the nitrogen content of fertilizers. There are many methods available to do this analysis. These methods are compared on the basis of measurability factors (LE 71).

The production of fertilizer in a continuous process sets the following numerical values:

$$T_a = 30 \text{ minutes} \qquad \sigma_x = 1.2\% \text{ nitrogen}$$
$$T_x = 66 \text{ minutes} \qquad m_a = 0.80\%$$

The methods of analysis are listed in Table 3.5 (LE 71). It is seen that

Table 3.5. Methods for Analysis of Nitrogen in Fertilizer[a]

Criterion and Analytical Technique	Dead Time of Analysis (minutes)	SD of of Analysis (% N)	$(m_d)_a$	m_n	m
Total N, classical distillation	75	0.17	0.32	0.99	0.24
Total N, DSM automated analyzer	12	0.25	0.84	0.97	0.65
NO_3–N, Technicon AutoAnalyzer	15.5	0.51	0.79	0.92	0.58
NO_3–N, specific electrode	10	0.76	0.86	0.85	0.58
$NH_4NO_3/CaCO_3$ ratio, X-ray diffraction	8	0.8	0.89	0.83	0.59
Total N, fast neutron activation analysis	5	0.17	0.93	0.99	0.74
Specific gravity, γ-ray absorption	1	0.64	0.98	0.88	0.69

Source: Reprinted with permission from Leemans (LE 71). Copyright (1971) American Chemical Society.

[a] A sampling frequency of 2 samples/hour is assumed, which means that $m_a = 0.80$; $(m_d)_a$ = measurability caused by analysis time; m_n = measurability caused by accuracy. The related process characteristics are mentioned in the text.

the highest m value is given by neutron activation. In a practical situation the cost of the equipment can play an important role in solution of a particular procedure. In static situations the analysis time becomes unimportant. Here the specific gravity measurement by γ-ray absorption, a very inaccurate method, gives the second-best results concerning information and is much cheaper than the neutron-activation method. When it is taken into account that the γ-ray absorption method can be repeated after 1 minute, ($T_a = 1$ minute), then for this method $m_{total} = 0.86$, by far the best method!

The influence of averaging the sample obtained during one sampling period is demonstrated as follows: in the process under investigation m_a equals 0.80. Sampling during 30 minutes ($= T_g$) between two consecutive analysis gives $m_g = \exp(-T_g/3T_x) = 0.86$ and $m_a \cdot m_g = 0.68$. The accompanying list of m values (Table 3.6) is obtained taking these measurability factors into account.

Table 3.6. Measurability of Nitrogen Analysis

Method	m_{total}
Devarda	0.20
AutoAnalyzer	0.50
Ion-specific electrode	0.50
X-ray diffraction	0.51
Automatic Devarda	0.56
γ-ray absorption	0.59
Neutron activation	0.64

3.2.6. Precision

Precision is one of the most used—and sometimes exclusively used—criteria for quality of an analytical method. As can be seen from the contents of this book, this is by no means justified. There are many quality criteria that are important in the description of an analytical method or procedure. However, precision has been and will be a very important yardstick when one is describing and evaluating an analytical method. Since there are many books on statistics—the field of science used to quantify precision—that describe the theory and practical applications (e.g., BE 75, DA 84, DI 69, MA 88b, MI 88, NA 63, SA 78, YO 75) in this book we give just a brief outline.

Precision expresses the closeness of agreement between repeated test results. As a quality parameter in analysis, precision can be described qualitatively as the quantity that is a measure for the dispersion of results when an analytical procedure is repeated on one sample. This dispersion of results may be caused by many sources. It is common practice in describing precision only to imply the sources that cause random fluctuations in the procedure. Because the analytical procedure furnishes results that are related to the composition of the sample, the scatter of the results will be around the expected value of the result if no bias exists. In analytical chemistry this scatter is more often than not of such a nature that it can be described as a normal distribution.

When the results are not normally distributed, a simple transformation often produces a normal distribution (Section 4.1.2). A normal distribution is the frequency distribution of samples from a population

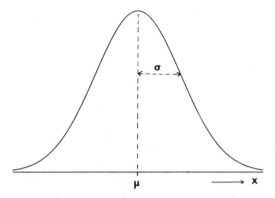

Figure 3.11. The normal probability function characterized by μ (mean) and σ (standard deviation).

that can be described by a Gaussian curve as shown in Figure 3.11. The normal, or Gaussian, distribution is characterized by the position of the mean, usually denoted as μ, and the half width of the bell-shaped curve at half height, proportional to the standard deviation σ. In practice only a limited number of observations are made and only estimates of μ and σ are obtained, usually denoted as $\bar{x}$ and s. Often estimates of s can be obtained from the range of the observations. The precision of an analytical method is often defined as the standard deviation s of a series of results obtained from one sample. The probability that a standard deviation s_1 is not significantly different from a standard deviation s_2, obtained from another series of observations on the same sample, should be established by statistical tests. For non-Gaussian distributions of the results, the common tests are not applicable. Therefore the distribution must be tested for normality (Section 4.1.1) and, if it is proved not to be normal, other tests should be applied, or the results must be transformed in such a way that a normal distribution results. In order to distinguish the various sources of scatter, an analysis of variance can be applied (Section 4.6).

The mean is only one type of measure of location of a distribution. Other characteristic quantities are the median and the mode. Since these measures are not often used in analytical quality control, we do not deal with their properties here. The property called simply the "mean" is in fact the arithmetic mean. Other types of mean are the geometric mean and the harmonic mean. The arithmetic mean for n

observations is defined as

$$\bar{x} = \frac{1}{n} \sum_{i=1}^{n} x_i \tag{3.33}$$

For random scatter in the results $\bar{x}$ is also the best available estimate of the population mean μ. As n increases, $\bar{x}$ becomes an increasingly precise estimate of μ. The computation of the mean is straightforward. However, because in the past the calculations often had to be done without calculators or computers, many devices have been used to reduce labor and the chance of errors. Some of the devices that have survived are the following:

1. When the observations consist of several digits, they can often be converted to smaller numbers by subtracting from each a constant quantity. The number of figures that have to be processed is reduced by this method. If the constant subtracted is x_0, then

$$\bar{x} = x_0 + \frac{1}{n} \sum_{i=1}^{n} (x_i - x_0) \tag{3.34}$$

2. When the observations all end in one or more zeros or are decimal fractions, they may be divided or multiplied by a power of 10, h:

$$\bar{x} = \frac{1}{hn} \sum_{i=1}^{n} (hx_i) \tag{3.35}$$

3. If a number of observations are available for each of which the arithmetic mean has already been calculated, the grand mean can be obtained by

$$\bar{x} = \frac{1}{\sum\limits_{i=1}^{k} n_i} \sum_{i=1}^{k} n_i \bar{x}_i \tag{3.36}$$

where the ith group of observations contains n_i observations with group mean $\bar{x}_i$. For equal sample sizes this reduces to

$$\bar{x} = \frac{1}{k} \sum_{i=1}^{k} \bar{x}_i \tag{3.37}$$

The geometric mean is defined as

$$\bar{x}_G = \left(\prod_{i=1}^{n} x_i \right)^{1/n} \tag{3.38}$$

or

$$\log \bar{x}_G = \frac{1}{n} \sum_{i=1}^{n} x_i^{-1} \tag{3.39}$$

that is, the nth root of the product of n observations.

The principal application is in averaging a sequence of ratios. The use of the geometric mean thus amounts to a transformation of the variable to $\log x_i$. As a rule it is better to stick to the transformation and refer to all characteristics as transformed characteristics instead of using a separate name, that is, $\bar{x}_G$.

The harmonic mean is defined by

$$\bar{x}_H = \frac{1}{n} \sum_{i=1}^{n} x_i^{-1} \tag{3.40}$$

This is equivalent to a transformation to the inverse of the variables.

An important descriptor of dispersion is the variance or its square root, the standard deviation. The variance of a population is the mean squard deviation of the individual values from the population mean and is denoted by σ^2. When the variance is estimated from a finite set of data, the symbols var or s^2 are used. The standard deviation s, the positive square root of the variance, has the same dimension as the data and therefore as a rule is used in the final result.

In computations, however, it is often more relevant to use the variance:

$$\sigma^2 = \frac{1}{n} \sum_{i=1}^{n} (x_i - \mu)^2 \qquad (n \to \infty) \tag{3.41}$$

The estimate s^2 of σ^2 is

$$s^2 = \frac{1}{n-1} \sum_{i=1}^{n} (x_i - \bar{x})^2 \qquad (3.42)$$

An often-used relationship that can be derived from eq. 3.42 is

$$s^2 = \frac{\sum_{i=1}^{n} x_i^2 - \left(\sum_{i=1}^{n} x_i\right)^2 \Big/ n}{n-1} \qquad (3.43)$$

The denominator $n-1$ represents the number of independent observations when the grand total of the results, $\sum x_i$, is used, for one of the observations can be calculated when all other results and $\sum x_i$ are known. The term $n-1$ is called the number of degrees of freedom.

Repeatability and reproducibility are subsets of the term precision. *Repeatability* is the measure of variations between test results of successive tests of the same sample carried out under the same conditions, i.e., the same test method, the same operator, the same testing equipment, the same laboratory, and a short interval of time. *Reproducibility* is the term used to express the measure of variations between test results obtained with the same test method on identical test samples under different conditions, i.e., different operators, different testing equipment, different laboratories, and/or different time.

A measure of dispersion that has no theoretical value but is sometimes used as a quality criterion is the coefficient of variation, or relative standard deviation (RSD), the standard deviation expressed as a percentage of the arithmetic mean (no symbol for this criterion is in common use):

$$\text{RSD} = \text{coefficient of variation} = 100 \, s/\bar{x} \qquad (3.44)$$

3.2.7. Accuracy

The concept of accuracy is one of the most difficult topics in analytical chemistry. Not only is the definition vague and difficult to interpret, but also the theoretical background of the methods for computing accuracy and the methods for estimating accuracy are complicated, ambiguous, and not generally accepted (AS 71).

Some of the definitions encountered in analytical chemical literature are given here. In a paper in *Analytical Chemistry* on statistical term it is stated (AN 75):

> Accuracy normally refers to the difference (error or bias) between the mean, $\bar{x}$, of the set of results and the value $\hat{x}$, which is accepted as the true or correct value for the quantity measured. It is also used as the difference between an individual value x_i and $\hat{x}$. The absolute accuracy of the mean is given by $\bar{x} - \hat{x}$ and of an individual value by $x_i - \hat{x}$.

The commission on analytical nomenclature of IUPAC (FE 69) used the word *bias* to denote accuracy:

> BIAS. The mean of the differences, having regard to sign, of the results from the true value. This equals the difference between the mean of a series of results and the true value.

The commission on spectrochemical and other optical procedures for analysis of IUPAC (AN 76) stated:

> *Accuracy* relates the agreement between the measured concentration and the "*true value*". The principal limitations on accuracy are: (a) random errors; (b) systematic errors due to *bias* in a given analytical procedure; bias represents the positive or negative deviation of the mean analytical result from the known or assumed true value; and (c) in multicomponent systems of elements, the treatment of interelement effects may involve some degree of approximation that leads to reproducible but incorrect estimates of concentrations.

Cali (CA 75) stated:

> An accurate measurement is one that is both free of systematic error and precise. Three major components of the measurement process must be attended to:
>
> - an agreed-on system of measurement
> - methods of demonstrated accuracy
> - reference materials

In our opinion "accuracy" cannot define a quantity; it only refers to the degree of attainability of the theoretical concept of "accurate." An accurate measurement is one that is both free of bias and precise. Bias is the mean of the differences of the results from the known or assumed true value. A precise measurement is one that shows no scatter in the results when repeated.

The difficulty in establishing a common definition of accuracy is illustrated by the many new measures that are proposed. To mention some of them:

Eckschlager and Štěpánek (EC 80) defined a measure of *negentropy* (Section 4.2):

$$A(p|q) = ld \frac{1}{\sigma\sqrt{2\pi}} - \frac{1}{2}\left(\frac{\delta}{\sigma}\right)^2 \tag{3.45}$$

where　$A(p|q)$ is the negative value of the information $H(p|q)$ gained by going from an a priori distribution $q(x)$ to an a posteriori distribution $p(x)$
ld is $\log_2$
$\delta = \mu - \mu_0$ (bias)
$\sigma = $ precision

McFarren et al. (MF 70) defined *total error* as

$$T = \frac{\text{abs}(\delta) + 2\sigma}{\mu_0} \cdot 100\% \tag{3.46}$$

A method with a total error less than 25% was qualified as excellent; 25–50% was considered acceptable. This method was further refined by Midgley (MI 77).

Westgard et al. (WE 74) defined *total analytical error*, a rather elaborate method.

Wolters et al. (WO 88) defined the *maximum total error*, the maximum difference between a measured value and the true value that can occur with a probability of 95%:

$$\text{MTE}(u) = \text{RE}(u) + \max\{\text{abs}[\text{SE}(u)], \text{abs}[\text{SE}(l)]\} \tag{3.47}$$

where　$\text{MTE}(u) = $ upper limit maximum total error
　　　　$\text{RE}(u) = $ upper limit random error, comprising the χ^2 distribution
　$\text{SE}(u)$ or $\text{SE}(l) = $ upper (lower) limit of the systematic error

Although accuracy seemingly cannot be quantified without elaborate calculations, it is possible to measure some properties that are related to the concept of accuracy. Precision can be measured, although a description of the circumstances of the measurement is required (Section 3.2.6). Bias can be estimated in some ways. In analytical chemistry often the only requirement for a result is that it be comparable with other results. An exception is when the true value must be known. Therefore, as a rule results of analysis are compared with the results obtained from the analysis of "standards" or "reference materials," i.e., materials with known or assumed properties. The property of the standard to be known, $\hat{x}$, can be obtained in various ways.

1. *Primary standards.* The theoretical composition of material of high purity can be used. The purity must be ascertained by independent methods, for example, by melting or boiling points, X-ray diffraction, and electrical properties. These types of standards are used in spectroscopy and titrimetry.

2. *Secondary standards.* The composition of these materials is measured by agreed-on methods by institutes equal to this task. A well-known institute in the United States that submits certified material is the National Institute of Standards and Technology (NIST), formerly the National Bureau of Standards (NBS) (CA 75). In Europe the Bureau Communautaire de Référence in Brussels fulfills this task. Where possible, the NIST certifies the numerical value of the property(ies) under investigation as "accurate"; that is, within stated uncertainty they are "true values." This is accomplished by three alternative methods:

a. Measurement by a method of known and demonstrated accuracy performed by two or more analysts working independently.

b. Measurement by two or more independent and reliable methods whose estimated inaccuracies are small, relative to accuracy required for certification.

c. Measurement via a qualified network of laboratories.

3. *Standard methods.* The composition of the material to be known can be obtained by applying an agreed-on method of analysis, for instance, a method issued by the International Standardization Organization (ISO) and the American Society for Testing Materials (ASTM). The value obtained by analysis according to a standard method can be assumed to be a "true value." This method of ensuring accuracy is often used in trade.

4. *Mean is true.* The mean of the results of a number of independent. selected laboratories can be assumed to be the "true value." It must be ascertained that all participants use comparable methods of data presentation and data handling (YO 75).

When the degree of accuracy is estimated, two related questions are often encountered:

1. When is the bias significant, or what is the probability that a found bias is real?
2. How can bias be distinguished from random errors?

The latter problem is dealt with in the discussions of analysis of variance (Section 4.6) and sequential analysis (Section 3.5). It should be stressed that random errors may seem to be bias, in particular when the random error is large compared to the bias. If a sample that is too small is taken from an inhomogeneous object, for example, the result of the analysis may deviate appreciably from the results of a thoroughly mixed standard. Repeating the analysis with a standard method may indicate that no real bias is present. Actually here the limited number of samples has caused the apparently significant bias. Even when the analytical method shows no bias at all, the method cannot be designated as accurate when the precision is insufficient. This also implies that it is useless to look for an accurate method of analysis in situations where samplig shows a bias or random error that is large compared to bias and precision of the method of analysis.

The Student's-*t* test can be applied to test the significance of a detected difference between the measurement results of sample and standard. This method is based on the assumption that the mean value of a series of observations shows a distribution around the mean value estimated from an infinitely large series, just as individual observations show a distribution around their mean. The distribution of the means has a probability density function that resembles the well-known Gaussian distribution for individual observations. This distribution is called the *t* distribution. Values of *t* are tabulated (see Table 3.7) for the probability that a particular mean belongs to a given population of means. If we state as hypotheses

H_0: $\bar{x} = \mu_0$ with $\bar{x}$ = mean of tested population
H_1: $\bar{x} \neq \mu_0$ with μ_0 = true mean of a given total population

Table 3.7. Probability Points of the t Distribution (Single-Sided)

ϕ	P				
	0.1	0.05	0.025	0.01	0.005
1	3.08	6.31	12.70	31.80	63.70
2	1.89	2.92	4.30	6.96	9.92
3	1.64	2.35	3.18	4.54	5.84
4	1.53	2.13	2.78	3.75	4.60
5	1.48	2.01	2.57	3.36	4.03
6	1.44	1.94	2.45	3.14	3.71
7	1.42	1.89	2.36	3.00	3.50
8	1.40	1.86	2.31	2.90	3.36
9	1.38	1.83	2.26	2.82	3.25
10	1.37	1.81	2.23	2.76	3.17
11	1.36	1.80	2.20	2.72	3.11
12	1.36	1.78	2.18	2.68	3.05
13	1.35	1.77	2.16	2.65	3.01
14	1.34	1.76	2.14	2.62	2.98
15	1.34	1.75	2.13	2.60	2.95
16	1.34	1.75	2.12	2.58	2.92
17	1.33	1.74	2.11	2.57	2.90
18	1.33	1.73	2.10	2.55	2.88
19	1.33	1.73	2.09	2.54	2.86
20	1.32	1.72	2.09	2.53	2.85
21	1.32	1.72	2.08	2.52	2.83
22	1.32	1.72	2.07	2.51	2.82
23	1.32	1.71	2.07	2.50	2.81
24	1.32	1.71	2.06	2.49	2.80
25	1.32	1.71	2.06	2.48	2.79
26	1.32	1.71	2.06	2.48	2.78
27	1.31	1.70	2.05	2.47	2.77
28	1.31	1.70	2.05	2.47	2.76
29	1.31	1.70	2.05	2.46	2.76
30	1.31	1.70	2.04	2.46	2.75
40	1.30	1.68	2.02	2.42	2.70
60	1.30	1.67	2.00	2.39	2.66
120	1.29	1.66	1.98	2.36	2.62
∞	1.28	1.64	1.96	2.33	2.58

then, if H_0 is true,

$$t\left(\frac{\alpha}{2}, n-1\right) \leqslant \frac{\bar{x} - \mu_0}{s/\sqrt{n}} \leqslant t\left(\frac{1-\alpha}{2}, n-1\right) \qquad (3.48)$$

where $n - 1 =$ degrees of freedom
$s =$ standard deviation
$\alpha =$ significance level

or

$$t_{(\alpha/2, n-1)} \sqrt{s^2/n} \leqslant \bar{x} - \mu_0 \leqslant t_{[(1-\alpha)/2, n-1]} \sqrt{s^2/n} \qquad (3.49)$$

Equation 3.49 can be applied only when μ_0 is known. If μ_0 has to be estimated with a finite number of observations (e.g., $\bar{x}_2$ from n_2 observations), then

$$t_{(\alpha/2, n_1+n_2-2)} \left[\left(\frac{1}{n_1} + \frac{1}{n_2}\right)s^2\right]^{1/2}$$

$$\leqslant \bar{x}_1 - \bar{x}_2 \leqslant t_{[(1-\alpha)/2, n_1+n_2-2]} \left[\left(\frac{1}{n_1} + \frac{1}{n_2}\right)s^2\right]^{1/2} \qquad (3.50)$$

As is obvious, the proof that $\bar{x}_1$ and $\bar{x}_2$ differ with a certain significance does not prove that the method with $\bar{x}_1$ is not accurate, even if $\bar{x}_2$ has been estimated from a large number of observations; $\bar{x}_2$ is just the mean of observations. Only if it has been ascertained by many means that $\bar{x}_2$ is the true mean (μ_0) disagreement between $\bar{x}_1$ and $\bar{x}_2$ tells us something about accuracy.

3.2.8. Measuring Functions and Calibration Functions

During the analysis of a sample, a number of actions are performed, resulting in one or more measured data that are used to calculate quantitative analytical results. From this point of view analysis may be considered to be a process with input parameters on composition, denoted as concentrations x_i and with output parameters measured data y_j:

$$x_i \rightarrow \text{PROCEDURE} \rightarrow y_j$$

where $i = 1, 2, \cdots, n$
        $j = 1, 2, \cdots, m$

In order to evaluate from y_j the value for x_i, the functional relationship between y_j and x_i should be known. The establishment of this functional relationship is called calibration. An analytical method for analysis of n compounds $x_1, x_2, \ldots, x_n$ of a sample should produce $y_1, y_2, \ldots, y_m$ measured data $(m \geqslant n)$.

These data should be mutually independent, as are the concentrations of the n compounds. The functional relationship between measured data y and composition x is called the calibration function. In order to apply a given analytical procedure for chemical analysis, each measurement y_j should correspond to a composition x_i according to

$$x_i = y_j \qquad (y_j, \text{ all } x \text{ except } x_i) \tag{3.51}$$

In general this condition implies that in order to measure one concentration value x_1, all other n concentrations $x_j (j \neq i)$ should be known, or that these values should be measured by collecting at least $n - 1$ more y_i values. This implies for quantitative analysis the solution of m equations with n unknowns $(m \geqslant n)$:

$$
\begin{aligned}
x_1 &= g_1 \qquad (y_1, y_2, \ldots, y_m) \\
x_2 &= g_2 \qquad (y_1, y_2, \ldots, y_m) \\
&\ \ \vdots \\
x_n &= g_n \qquad (y_1, y_2, \ldots, y_m)
\end{aligned}
\tag{3.52}
$$

These functions are named the analysis or measuring functions G: the calibration functions are $y = F \cdot x$. Here F contains no y terms and G contains no x terms. In general, the g functions are summations of y values, multiplied with weight functions μ, and the f functions are summations of x values, multiplied by weight functions, say, v. In linear situations, which are the only ones that can be handled in practice, at most $m \cdot n$ weighting coefficients should be known. Each of the components n should be used for calibrating, thus producing m y-values per component.

In order to solve for n unknowns only n independent equations are required, which means that out of the m measurements, $m - n$ can be eliminated. The number of calibration samples thus remains n, but the number of calibration measurements may diminish to $n - y$ values. In nonlinear situations in general there will be no solution.

This leaves the problem of what rows and what columns should be removed out of G and F. As a rule of thumb, one may state that the resulting $n \cdot n$ matrix should be such that its determinant has a maximal value. This means that the determinant has highest possible elements on the diagonal and smallest possible elements on the nondiagonal positions. A generally accepted method to reduce the number of columns is to multiply the matrices with their transpose (rows and columns interchanged).

This method is illustrated in the following example: in a spectrophotometric measurement on chlorine and bromine dissolved in a suitable solvent, extinction coefficients a_x are known and the extinctions obtained are shown in Table 3.8. In a mixture of Cl_2 and Br_2, one measures with an optical path length d (cm):

$$y_\lambda = \left(\log \frac{I_0}{I} \right)_\lambda = [C_{Cl_2} \cdot a_{Cl_2}(\lambda) + C_{Br_2} \cdot a_{Br_2}(\lambda)] \cdot d \qquad (3.53)$$

In order to reduce the six equations with two unknowns (the concentrations C_{Cl_2} and C_{Br_2}, we multiply the left-hand side as well

Table 3.8. Extinction Coefficients of Chlorine and Bromine

λ (nm)	a_{Cl_2} (1/mol·cm)	a_{Br_2} (1/mol·cm)	$\log(I_0/I)_\lambda = y_\lambda$ Mixture (optical path length 1 cm)
455	4.5	168	0.1995
417	8.4	211	0.2698
385	20	158	0.2980
357	56	30	0.4220
333	100	4.7	0.7047
312	71	5.3	0.5023

as the right-hand side with the transpose a^t of the extinction matrix a (see also Section 4.3.2, esp. 4.3.2.3).

$$a^t \cdot y_\lambda = a^t \cdot a \cdot x \cdot d \tag{3.54}$$

which becomes

$$
\begin{vmatrix}
4.5 & 8.4 & 20. & 56. & 100. & 71. \\
168. & 211. & 158. & 30. & 4.7 & 5.3
\end{vmatrix}
\cdot
\begin{vmatrix}
0.1995 \\
0.2698 \\
0.2980 \\
0.4220 \\
0.7047 \\
0.5023
\end{vmatrix}
=
$$

$$
\begin{vmatrix}
4.5 & 8.4 & 20. & 56. & 100. & 71. \\
168. & 211. & 158. & 30. & 4.7 & 5.3
\end{vmatrix}
\cdot
\begin{vmatrix}
4.5 & 168. \\
8.4 & 211. \\
20. & 158. \\
56. & 30. \\
100. & 4.7 \\
71. & 5.3
\end{vmatrix}
\cdot
\begin{vmatrix}
C_{\text{Cl}_2} \\
C_{\text{br}_2}
\end{vmatrix}
\tag{3.55}
$$

$$
\begin{vmatrix}
138.889370 \\
156.162080
\end{vmatrix}
=
\begin{vmatrix}
18667.81 & 8214.7 \\
8214.7 & 98659.18
\end{vmatrix}
\cdot
\begin{vmatrix}
C_{\text{Cl}_2} \\
C_{\text{Br}_2}
\end{vmatrix}
\tag{3.56}
$$

It is seen that the determinant F is well conditioned because $8214.7 < 18667.81$ and < 98659.18.

Application of Cramer's rule (provided det $\neq 0$) gives

$$C_{\text{Cl}_2} = \frac{F^{1,1}}{\|F\|} \cdot x \frac{F_{2,2} \cdot x_1 - F_{1,2} \cdot x_2}{F_{1,1} \cdot F_{2,2} - F_{1,2} \cdot F_{2,1}} = 0.0070 (\text{mol Cl}_2/\text{L}) \tag{3.57}$$

and

$$C_{\text{Br}_2} = \frac{F^{2,1}}{\|F\|} \cdot x \frac{-F_{1,2} \cdot x_1 + F_{1,1} \cdot x_2}{F_{1,1} \cdot F_{2,2} - F_{1,2} \cdot F_{2,1}} = 0.010 (\text{mol Br}_2/\text{L}) \tag{3.58}$$

in which $F = a^t \cdot a \cdot d$.

3.2.9.   Selectivity and Specificity

In quantitative analysis the measurement of one particular component of a mixture may be disturbed by other components of the mixture. This means that the measurement is nonspecific for the compound under investigation.

With a completely specific method for a compound the concentration of that component can be measured, regardless of what other compounds are present in the sample. The other components do not produce an analytical signal. A method of analysis is called completely selective when it produces correct analytical results for various components of a mixture without any mutual interaction of the components. A selective method thus is composed of a series of specific measurements. In situations where the linear system holds,

$$y = F \cdot x \qquad (3.59)$$

the measurement for component k is specific when only the terms $f_{i,k}$ $(i = 1, \ldots, n)$ are nonzero. Here the matrix is called F_{spec} and the value of y is a function of x_k only.

For a completely selective method each row of matrix F contains only one nonzero element. By interchanging rows, one may get a matrix F with all nonzero elements on the diagonal. This matrix is called F_{sel}. In practical situations F_{spec} and also F_{sel} contain other nonzero elements. These elements, however, have considerably smaller numerical values than the corresponding $f_{i,k}$ values. The undesired nonzero elements of F_{sel} and F_{spec} can be used for a quantitative description of selectivity and specificity (KA 73). Kaiser proposes a method for calculation of selectivity based on the concept that a procedure becomes more selective with a relative decrease of the nondiagonal elements with respect to the diagonal ones.

3.3.   SELECTION OF AN ANALYTICAL PROCEDURE

One of the difficulties that confronts an analytical chemist when a particular problem has to be solved is how to select a procedure out of the growing number of analytical methods available. During recent years the number of methods and their applications has increased to such an extent that it seems impossible to familiarize oneself with all possibilities by one's own experience or by study. In common practice this situation results in applying exclusively those methods and procedures that are familiar and well known. Apart from this, one should be aware of the simplification of one's choice because of the limitations of the instrumentation of a particular laboratory. Another aspect of the problem is that during a chemistry course it is not possible for a student to receive a complete and lucid picture of all existing analytical techniques and their application. These problems are well realized among analytical chemists, as may be concluded from a quotation of Kaiser (KA 70):

At present there is no systematic approach to answer the question: which one may be the best analytical procedure to solve an analytical problem. We are relying on the knowledge and experience of the analyst, we are scanning the literature, we are dependent on habits and fashions. Isn't that muddling through?

Therefore one should ask oneself, "Is it possible to simplify the selection of a method for solving a particular problem according to objective rules?" In order to formulate a set of objective rules one may start by listing criteria for analytical procedures such as limit of detection, selectivity, and accuracy. These criteria can be used for mapping all available procedures in a multidimensional space, where the number of criteria equals the number of dimensions of that space. By varying a procedure a particular analytical technique such as atomic absorption may cover a range of a certain criterion such as the detection limit. In the multidimensional methods representation space this results in a multidimensional plane or volume enclosing an area in which that technique can be successfully applied.

Sometimes different techniques are represented by areas that partly overlap, which means that in the overlapping area both techniques can be applied.

The same kind of representation may be developed for a problem space, in which each analytical problem, described by a set of problem criteria such as chemical compound analyzed for, required accuracy, and expected grade, is positioned. By mapping the methods space on the problem space one may see whether the problem is positioned within the boundaries of the method and conclude what method or methods are applicable. Tests should be made as to whether methods selected according to this procedure prove to be better than methods selected randomly.

This "multicriteria" problem is a very difficult one. Many authors have tried to define a method that circumvents the problem of weighting the different criteria. One of the most widely used is so-called Pareto optimization. With this method a number of combinations of criteria is given, each giving the maximum "yield" for the combination of all other criteria, if one criterium has been chosen. Keller, Massart and Brans (KE 91) described a case study of an analytical application. Other proposals are the ELECTRE (RO 72) and the PROMETHEE method (BR 89) and the use of a neural network as a learning machine (KA 90). A totally different approach is to select a method in a restricted domain by using all available knowledge about the problem under consideration. This knowledge can be collected and made available in a (series of) expert systems. Wellmann et al. (WE 86) described a decision system for the optimal selection of laboratory procedures that consists of an expert system using information extracted from literature. In a first step all method parameters are considered equal and are used to find a method that suits the desired parameters. When no appropriate answer can be found, more and more parameters are eliminated in inverse proportion to their significance. The program is set up for a very narrow domain, the determination of manganese and tungsten.

3.3.1.   Cost of Analysis

A selection criterion that requires special attention is the cost of a particular analysis. In order to estimate the cost C of an analysis one may use the following equation:

$$C = \tau_a f + 1.74 \times 10^{-6} \cdot F \cdot \tau_n \qquad (3.60)$$

where the first term on the right-hand side denotes the operating cost per analysis based on labor and laboratory overhead costs expressed in money per time unit f, and τ_a, the handling and preprocessing time by laboratory personnel, is expressed in the same time units (e.g., minutes per analysis). The second term on the right denotes the capital investment in analytical equipment F, reduced by the number of analyses performed during its economic life with a "normal" work load. This

Table 3.9. Prices of Instruments

Instrument Class	Investment (arbitrary units)									
	AAS	UV/vis	Tit	XRF	AES	GLC	FES	MS	IR	PMR
1	5	2	0	—	—	10	5	50	15	—
2	10	10	2.5	100	30	15	10	100	25	—
3	20	20	5	150	60	20	20	200	35	100
4	30	40	7.5	200	120	30	30	350	60	200
5	50	80	10	300	—	50	50	500	100	300

amount of money is multiplied by τ_n, the duration of one analysis, expressed in the proper time units (e.g., minutes). For an on-line analytical instrument, the number of analyses performed in its economic life is fixed and may be calculated from the frequency with which samples are analyzed, for example, for process control. The number of analyses performed with a laboratory instrument is not quite fixed because it depends on the number of analyses required. Here F should be estimated. It may be interesting to see how total cost C is composed for a value of f of 80 money units/hour and an investment of 10^5 money units when τ_a equals 30 minutes and τ_n 15 minutes:

$$C = 40 + 2.03 = 42.03 \tag{3.61}$$

which shows labor to be much more expensive than investments.

The cost of investment is the capital investment in equipment. This amount of money mainly depends on the quality of the instrument under consideration. As a yardstick Table 3.9 is used, based on five classes of instruments, ranging from primitive (1) through "normal" to advanced (5).

3.4. QUALITY CONTROL

In analytical work that has to be controlled, two levels can be distinguished. The first level pertains to the analytical procedure. Most of the quality parameters described earlier in this book are on the procedural level, for example, sampling size, sampling frequency, limit of detection, sensitivity, and precision. The second level belongs to the laboratory. When an analytical procedure is performed, not only are the results of various workers and instruments compared, but also the results of other laboratories where comparable procedures are applied.

In this section some methods of controlling the quality, based on the axioms mentioned, are described.

3.4.1. The Ruggedness Test

Method validation is the final stage in the development of an analytical method. It involves the assessment of different quality parameters, such as selectivity, accuracy, detection limits, and precision of the method (see Table 3.10). Precision involves the determination of the random error under specific circumstances.

The level at which a method has to be tested depends on its later use. Whenever an analytical method has to be applied in different sites or laboratories, eventually an interlaboratory test must be carried out to ensure that the method performs adequately in different circumstances. The variability of a method under identical circumstances, e.g., in the developer laboratory, is called the *repeatability*, whereas the precision in different circumstances is called the *reproducibility*. An interlaboratory study can be very costly and should only be undertaken when there is a reasonable chance that the method will be accepted. It is not an unusual situation that a method performs perfectly well in one laboratory and fails when tested in an interlaboratory study. This may be because the initiating laboratory tested the performance of the method under constant conditions. In one laboratory a nominal 1 M solution may be actually 0.95 M, whereas in another laboratory this is 1.02 M. To avoid problems in an interlaboratory test the influence of

Table 3.10. Tests in a Validation Procedure

Method of Validation	Tests for
Specificity	Interferences
Accuracy	Recovery Linearity
Sensitivity	Detection limit
Precision	Repeatability Ruggedness Reproducibility

Table 3.11. Ruggedness Test

1. Identification of critical factors
2. Planning of experiments
3. Experimental phase
4. Interpretation of result

this kind of variation of the operating conditions on the method's performance should be tested. The only way to assess this is to deliberately introduce changes in the operating conditions. These changes should be such that they reflect as well as possible the differences in operating conditions that are to be expected for the intended use of the method. The intended use determines the scale of the introduced variations. Such a study is called a *ruggedness test*. This test involves a number of essential steps (Table 3.11).

The first step is the identification of the critical factors that are expected to influence the method's performance. For an HPLC experiment a number of such factors is shown in Table 3.12. Together with the identifications, the scale of the variations that are to be tested must be defined. This too depends on the eventual intended use of the method. This step requires a thorough knowledge and experience of the analytical method.

Once the critical factors and their levels at which they should be tested are identified, an experimental strategy should be developed. The use of experimental design for ruggedness testing has been proposed by Youden and Steiner (YO 75) to evaluate the ruggedness of a method for certain factors. More recently these designs have been applied in practice for the ruggedness testing of HPLC methods. The number of factors that have to be tested is usually large. Even when only the most important factors are selected, a large number of experiments is still required. Experimental designs that enable one to extract as much information as possible from a minimum number of experiments are necessary. The basic idea is not to change one factor at a time but rather simultaneously to change several factors in such a way that it is possible to identify the changes that produce an influence on the method's performance (Figure 3.12).

Factorial designs are the most suitable for ruggedness testing. They enable one to establish the influence of each of the factors separately

Table 3.12. Example of Possible Critical Factors and
Levels for HPLC Experiments

Critical Factor	Deviation from Nominal Value
Sample preparation factors:	
Sample weight	1%
Shake time	20%
Sonicate time	20%
Heating temperature	5° C
Wash volume	30%
Extraction volume	30%
Centrifuge time	20%
Pore size of filter	5 μm
⋮	
Chromatographic factors:	
pH	1
Temperature	5 °C
Solvent %	3%
Flow rate	0.1 mL/min
Buffer concentration	1%
Additive concentration	0.5%
⋮	
Column factors:	
Manufacture	
Batch	
⋮	
Detector factors:	
Wavelength	5 nm
Time constant	5
⋮	

(main effects) and the interactions between them. An example of a factorial design is shown in Figure 3.12.

The number of experiments required is 2^n, n being the number of factors to be tested. For 7 factors this means $2^7 = 128$ experiments. In practice full factorial designs can only be used when the number of factors to be used is low. When it can be assumed that there are no significant higher order interactions between the factors, which is often a realistic assumption, the fractional factorial designs are most efficient.

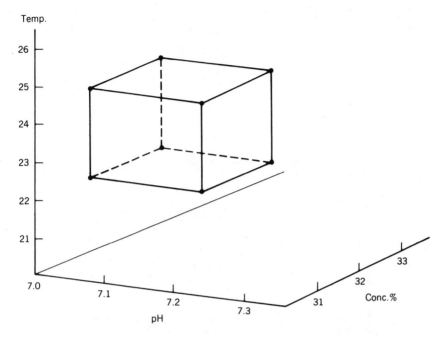

Figure 3.12. A two-level three-factor factorial design.

Saturated fractional factorial designs enable one to test $N - 1$ factors with N experiments. These designs are very efficient and labor conservative, but they do not anymore allow one to establish interactions between the factors. Plackett and Burman (PL 46) have produced these designs for values of N that are a multiple of 4. An example of a Plackett–Burman design for 12 experiments to test a maximum of 11 factors is shown in Table 3.13. When the critical factors are identified and the experimental design is selected, the experimental work can be carried out. This involves performing the experiments prescribed by the experimental design. To avoid artifacts, it is necessary to perform these experiments in a random sequence. The performance of the methods for the different experiments must be recorded. For HPLC experiments several performance parameters can be selected. The most relevant of these are the following: the concentration calculated from peak height or peak area; the retention time; the resolution, etc. In general it is important to measure as many parameters as possible to obtain a picture of the method that is complete as possible.

Table 3.13. Plackett–Burman Design

Expt.	Factor											4	Results
	I	II	III	IV	V	VI	VII	VIII	IX	X	XI		
	—	—	—	—	—	—	—	1	1	1	1	—	
	1	1	-1	1	1	1	-1	-1	-1	1	-1		
1	76	76	-76	76	76	76	-76	-76	-76	76	-76	76	76
2	12	-12	12	12	12	-12	-12	-12	12	-12	12	12	12
3	-21	21	21	21	-21	-21	-21	21	-21	21	21	21	21
4	25	25	25	-25	-25	-25	25	-25	25	25	-25	25	25
5	360	360	-360	-360	-360	360	360	360	360	-360	360	360	360
6	318	-318	-318	-318	318	-318	318	318	-318	318	318	318	318
7	-29	-29	-29	29	-29	29	29	-29	29	29	29	29	29
8	-314	-314	314	-314	314	314	-314	314	314	314	-314	314	314
9	-50	50	-50	50	50	-50	50	50	50	-50	-50	50	50
10	25	-25	25	25	-25	25	25	25	25	-25	-25	25	25
11	-30	30	30	-30	30	30	30	-30	-30	-30	30	30	30
12	-230	-230	-230	-230	-230	-230	-230	-230	-230	-230	-230	230	230

When the experiments are performed the effects of the different factors can be calculated. They are calculated from the differencs between the experiments that involved the factor at the high level and those that involved the factor at the low level. In Table 3.13 it can be seen that for the first factor the experiments 1, 2, 4, 5, 6, and 10 involve the factor at the extreme level, whereas the experiments 3, 7, 8, 9, 11, and 12 involve this factor at the nominal level. The main effect of factor 1 is then calculated as

$$ME(1) = \frac{Y1 + Y2 + Y4 + Y5 + Y6 + Y10}{6}$$

$$= \frac{Y3 + Y7 + Y8 + Y9 + Y11 + Y12}{6} \tag{3.62}$$

The main effects can be calculated in this way for each of the factors. They can now be listed in order of size. If no factor has an influence, all the effects should be small and of about the same size. If one or two factors have an influence on the performance of the method, then the values of their main effects will be much larger then the other main effects. Obviously a method should not be influenced by the small variations introduced for the ruggedness test that are expected to be encountered in practice. If there are no large differences, the analytical error can be estimated from the mean of the main effects:

$$SE = \left[\frac{2}{11} \sum_{i=1}^{11} ME(i)^2 \right]^{1/2} \tag{3.63}$$

The estimate of the standard deviation reflects the differences that have been created to simulate the situation that different laboratories have been carrying out the experiments. If this standard deviation is unsatisfactorily high, then a interlaboratory test will certainly yield unsatisfactory results.

If one or two factors show an important influence on the method's performance, then the method should be improved to make it more rugged to the small variations in these factors. When this is not possible, then the method procedure should specify a strict range of these factors, which is chosen so that the performance is adequate when the factor remains in that interval.

If only those methods that survived such a ruggedness test were subjected to an interlaboratory test, much disappointment would be avoided.

The method has been extended and implemented in an expert system described by van Leeuwen et al. (LE 91a, LE 91b) (Figure 3.13). In this system all factors are tested systematically as to whether they are

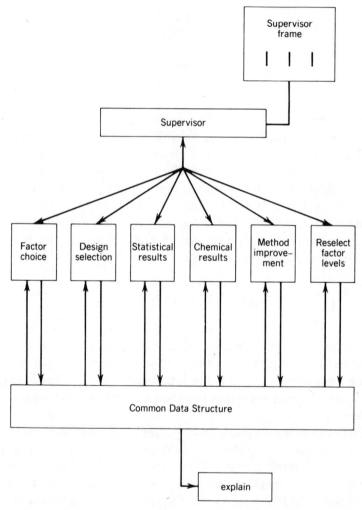

Figure 3.13. Structure of an expert system for ruggedness analysis.

statistically significant and their mutual influence is given. A complete setup of the experiments required to make a justification of the final advice is given. The usefulness of an expert system for that purpose lies in the objective advice it offers, taking care of all relevant interferences.

3.4.2. Control Charts

One of the oldest methods used to monitor and control quality of a procedure—in particular, the accuracy and precision—is the control chart (DA 84). A control chart is a graphic representation on which the values of the quality characteristic under investigation are plotted sequentially. The type of chart devised by Shewhart (SH 31) consists of a central line and two limit lines spaced above and below the central line. These are usually termed the inner and outer control limits. The distribution of the plotted values with respect to the control limits provides valuable statistical information on the quality characteristic being studied. Another type of chart shows the cumulative differences of the quality characteristic from a target level also plotted sequentially. This chart is known as a cumulative sum chart, or *cusum chart*. Of course a control chart is not an end in itself; it is an aid in prediction and a basis for action. The control chart may take a variety of forms. This section is concerned with the basic principles; detailed explanations and examples can be found in books on quality control of production processes (e.g., DA 84).

3.4.2.1. Shewhart Control Charts

If the chart is used for subgroup averages, the main steps to be taken are as follows. Use the individual readings of the property on which the quality control is based, for example, the composition of a standard sample analyzed according to the analytical method to be controlled. Plot the readings in the order in which they were obtained. Sometimes the observations can be subdivided into a number of rational subgroups within which there is reason to believe that only chance causes of variation have operated but between which assignable causes may have operated. Thus a subgroup may consist of the results emanating from one repeated experiment if the results are obtained within a short time span and by one person. If none of the plotted points falls outside the outer control limits, it can be concluded that the data are statisti-

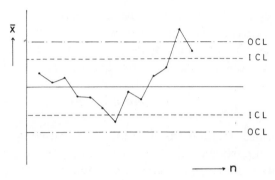

Figure 3.14. Control chart: OCL, outer control limit; ICL, inner control limit.

cally uniform, that is, no major assignable causes of variation have intervened during the period (Figure 3.14).

The drawing of the limiting lines for subgroup averages is based on the following principles. If individual observations from the same population, for example, results obtained from a standard sample, are distributed about a mean level μ with a standard deviation σ, then the averages of subgroups, each containing n individuals, are distributed around μ with a standard error of $\sigma/\sqrt{n}$. Whatever the distribution of the individuals, the distribution of subgroup averages approaches normality as n increases. Even when n is as small as 4 or 5, the distribution of averages is close to normal provided the distribution of the individuals is not extremely asymmetric. When subgroups are composed of readings obtained in experiments repeated within a short time span, it is not probable that the distribution is asymmetric. Provided that the data are statistically uniform, only one subgroup mean in 40 will on the average lie above the limit $\mu + 1.96\sigma/\sqrt{n}$ and 1 in 40 below the limit $\mu - 1.96\sigma/\sqrt{n}$, that is, 1 in 20 outside $\mu \pm 1.96\sigma/\sqrt{n}$. These limits are referred to as the upper and lower 1 in 40 (or 0.025) lines, or inner control lines, respectively.

Further, only one subgroup mean in 1000 will lie above the limit $\mu + 3.09\sigma/\sqrt{n}$ and 1 in 1000 below $\mu - 1.96\sigma/\sqrt{n}$, that is, 1 in 500 outside $\mu \pm 1.96\sigma/\sqrt{n}$. These limits are known as the 1 in 1000 (or 0.001) lines, or outer control lines.

To simplify the calculation and the verbal description in practice the multiplying factor 1.96 is rounded to 2 and 3.09 to 3. In some parts of the world, as in the United States, only the $3\sigma/\sqrt{n}$ limits are used.

Common practice is to base action on a single point falling beyond the outer limits or successive points outside the inner limit.

The justification for these choices of control limits to indicate a variation that should not be allowed to pass is that practical experience has shown that occurrence of points beyond them is a very useful signal for action to be taken. The correspondence of control limits to explicit probabilities derived from the normal distribution curve may be of value when one is interpreting the charts, but too much attention should not be paid to the precise probability values associated with the particular multiple of the standard error that is adopted. When quantitative results on quality are required, a more stringent scheme of calculations should be followed, for example, an analysis of variance scheme (Section 4.6) or sequential analysis (Section 3.5). The speed of discovery that something is out of control can be estimated with the following reasoning:

Suppose the average value suddenly shifts away from μ by an amount equivalent to $3\sigma/\sqrt{n}$. Then there is a 1 in 2 chance that the next subgroup average will fall beyond the outer control limit and initiate action. The chance of detection after two readings will be $\frac{1}{2} + (\frac{1}{2} \times \frac{1}{2})$, etc. The average number of tests before the change is detected is termed the average run length. For a $3\sigma/\sqrt{n}$ shift of the mean, the average run length is 2. In general the average run lengths is $1/A$, where A is the tail area of the normal distribution beyond the control line.

3.4.2.2. Chart for Subgroup Ranges

In this chart the quality parameter precision is plotted and used for control. Now the standard deviation s or the range R of a subgroup of results consisting of n observations is used to construct a control chart. The range R as a rule should be used when the subgroup size $n < 13$, as is to be expected in controlling an analytical procedure. The central line for a chart for subgroup ranges is σ_0 or s_0, the estimate of σ_0. The inner and outer limits will be fixed at $D_{0.025}\sigma$ and $D_{0.001}\sigma$, respectively. The same probability levels are used as in the chart for subgroup averages: 0.001 in each tail for the outer limits, and 0.025 in each tail for the inner limits.

Table 3.14. Outer and Inner Limits for Subgroup Ranges

No. of Observations in a Subgroup (n)	Lower Limit		Upper Limit	
	Outer Limit ($D_{0.001}$)	Inner Limit ($D_{0.025}$)	Inner Limit ($D_{0.975}$)	Outer Limit ($D_{0.999}$)
2	0.00	0.04	3.17	4.65
3	0.06	0.30	3.68	5.06
4	0.20	0.59	3.98	5.31
5	0.37	0.85	4.20	5.48
6	0.54	1.06	4.36	5.62
7	0.69	1.25	4.49	5.73
8	0.83	1.41	4.61	5.82
9	0.96	1.55	4.70	5.90
10	1.08	1.67	4.79	5.97
11	1.20	1.78	4.86	6.04
12	1.30	1.88	4.92	6.09

Note that the lower and upper limits are not symmetric about the average, because the chance that the range will be higher than the average is not equal to the chance that the range will be lower than the average (Table 3.14).

3.4.2.3. Cusum Charts

The charts for average and range can be used when sudden changes should be detected. In controlling the quality of analytical methods, this is particularly so when sudden changes in the method, application of wrong procedures, interchange of reagents, and the breakdown of parts of instruments have to be detected. When more gradual changes should be detected, such as aging of reagents and obsoleteness of standard curves, another method of plotting the results should be used. This method is known as the cusum chart.

Here the decision is based on the whole of the recent sequence of measurements. Instead of plotting the subgroup averages $\bar{x}_1, \bar{x}_2, \ldots,$ a reference value k is chosen and the cumulative sum is calculated:

$$s_i = \sum_{j=1}^{i} (\bar{x}_j - k) \tag{3.64}$$

Each sum s_i is obtained from its predecessor by the operation of adding the new difference; k is a reference value that is chosen equal to the average, but calculation of the cumulative sums (abbreviated to cusum) is simplified if a rounded value is chosen. If the mean value of the subgroup averages remains close to the reference value, some of the differences are positive and some negative, so that the cusum chart is essentially horizontal. However, if the average value of the process rises to a new constant level, more of the differences become positive and the mean path slopes upward. The position of the change in the cusum slope indicates the moment the change in the quality parameter occurred. However, the visual picture depends to some extend on the scales chosen for the axes of the chart. If the horizontal distance between the plotted points is regarded as one unit, it is recommended that the same distance on the vertical scale represent approximately $2\sigma/\sqrt{n}$ property units, where σ is the population standard deviation.

The unit distance on the vertical scale does not have to be exactly equivalent to $2\sigma/\sqrt{n}$ units, and in fact it eases the task of plotting the cusum if a suitable rounded lower value is chosen.

Just as action decisions from a Shewhart chart are signaled by points falling outside the control limits, a rule is needed to decide when the change of slope indicates some assignable cause in the method of analysis. One method for taking these decisions is to use a V-shaped mask. This is a graphic aid that can be used to quantify the disturbance in the cusum. The mask can be constructed as a solid truncated V in a material such as cardboard, or the V can be engraved in a sheet of transparent Perspex. The mask is positioned on the cusum with the axis of the V parallel to the horizontal axis of the chart (Figure 3.15). The point P, at a distance $d = 2t$ (the unit length of the horizontal axis of the cusum chart) from T, is superimposed on the most recent cusum value S_t. If one of the preceding cusum S_i's is positioned below the limit of the V, OT, this is an indication of a significant positive change of μ. If the cusum line crosses limb BT, μ has changed in a downward direction. In both circumstances action is undertaken. A V-mask with an angle $\Phi = 30°$ and a lead distance d as indicated is equivalent to $3\sigma/\sqrt{n}$ limits. This has an average run length of 500 points.

As the cusum technique requires regular updating, especially when a drift has been detected, other methods have emerged. A method that can be used continuously is Trigg's monitoring technique (CE 75). The method uses the exponentially weighted average (see Section 4.3.2,

ANALYSIS

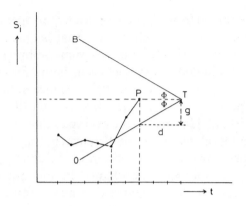

Figure 3.15. Cusum chart with mask.

esp. 4.3.2.2) to follow the observations.

$$S_t = \alpha x_t + (1 - \alpha)S_{t-1} \tag{3.65}$$

where α = factor to be chosen [0.1–0.2 in most cases, but a Kalman gain factor (Section 4.3.2, esp. 4.3.2.2) can be used as well]

x_t = last measurement

S_t = the best available prediction of the true value X_t

The prediction error is given by

$$\varepsilon_t = x_t - S_{t-1} \tag{3.66}$$

This error can be smoothed in the same way as S, so

$$\bar{\varepsilon}_t = \alpha \varepsilon_t + (1 - \alpha)\bar{\varepsilon}_{t-1} \tag{3.67}$$

Here $\bar{\varepsilon}_t$ is compared with the smoothed mean absolute deviation

$$\bar{e}_t = \alpha e_t + (1 - \alpha)\bar{e}_{t-1} \tag{3.68}$$

where $e_t = |\bar{\varepsilon}|$.

Now Trigg's tracking signal T_t is calculated by

$$T_t = \bar{\varepsilon}_t / \bar{e}_t \tag{3.69}$$

Table 3.15. Tracking Signal Values (CE 75)

Confidence Level (%)	Tracking Signal	
	$\alpha = 0.1$	$\alpha = 0.2$
70	0.24	0.33
80	0.29	0.40
85	0.32	0.45
90	0.35	0.50
95	0.42	0.58
98	0.48	0.66
99	0.53	0.71

The confidence level indicating the possibility of a trend is given in Table 3.15. As in all recursive methods a starting value should be given to obtain soon reliable results. Recommended is

$$\bar{\varepsilon}_{t-1} = 0; \qquad \bar{e}_{t-1} = (s)\sqrt{2/\pi}; \qquad S_{t-1} \simeq \sum_{\tau=-5}^{-1} x_{t-\tau} \qquad (3.70)$$

with s calculated from at least five measurements.

3.4.2.4.   Sequential Analysis

As might be supposed, sequential analysis (Section 3.5) can be used for quality control of analytical chemical procedures. In fact, the cusum technique is a somewhat simplified variant of sequential analysis. Therefore the cusum chart seems more suitable to the needs of control in a laboratory, for one of the axioms used states that all laboratory workers can evaluate the quality measurements and undertake action when required.

3.4.3.   Youden Plot

A very popular and effective controlling device for analytical laboratories is the round-robin test with graphic reporting. The method was described in detail by Youden (YO 72) and is therefore often called the *Youden plot method* (Youden himself refers to it as the two-sample chart). He demonstrates the role of systematic and random errors or, stated in another way, that of accuracy and precision. The data for

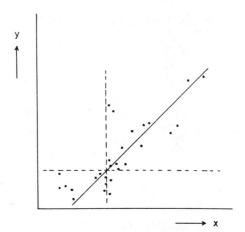

Figure 3.16. The Youden plot.

these charts (Figure 3.16) are obtained by having two similar materials, designated x and y, analyzed by each collaborator with a request for one determination per sample. Expanded rectangular x and y scales form the representation space for the pair of results provided by each laboratory. The pattern formed by the points conclusively demonstrates the major role played by the various systematic errors. Consider that random errors are really the cause of the scatter. In such a situation the two determinations may err in being both low, both high, x low and y high, or x high and y low. Since random errors are equally likely, all of the four possible results should be equally likely. Thus in every quadrant, formed by subdividing the field around the mean of the points in the representation space, the number of points corresponding to the laboratories should be equal. Otherwise stated, the probability density function consists of concentric circles with the mean as center. If the points are considered to have a normalized Gaussian probability function, the x results are scattered around the mean x_0 with a probability density

$$P_x = \frac{1}{\sigma\sqrt{2\pi}} \exp\left[-\frac{(x-x_0)^2}{2\sigma^2}\right] \qquad (\sigma = \text{standard deviation}) \quad (3.71)$$

The same situation holds for the y results. The standard deviation σ of x

and y are the same because the samples were similar; thus

$$P_y = \frac{1}{\sigma\sqrt{2\pi}} \exp\left[-\frac{(y-y_0)^2}{2\sigma^2}\right] \tag{3.72}$$

The probability density function of the points (x, y) in the representation space equals

$$P_{x,y} = P_x \cdot P_y = \frac{1}{2\pi\sigma^2} \exp\left\{-\frac{1}{2\sigma^2}[(x-x_0)^2 + (y-y_0)^2]\right\} \tag{3.73}$$

In Figure 3.17 the lines with equal density for σ and 2σ are shown. As can be seen from eq. 3.73 these lines are concentric circles. Frequently, however, no such distribution has been found. The points are nearly always found dominantly in two quadrants: the $++$ upper right quadrant and the $--$ lower left quadrant. Generally the points form an elliptical pattern with the major axis of the ellipse running diagonally at an angle of 45° to the x-axis, instead of being scattered within a circle. This indicates that, when a laboratory gets a result denoted as "high" (in reference to the consensus) with one material, it is most probable to be similarly high with the other material. The same statement holds for low results. The elongation of the ellipse is a measure for the accuracy of the method, scales along the x- and y-axes being the same. The distance of each point along the 45° axis to the mean is a measure of the systematic error that the laboratory is likely to have been making.

This is in complete agreement with the common understanding that, although good checks are obtained within one laboratory owing good

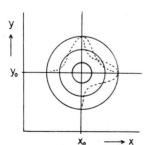

Figure 3.17. The probability density of the Youden plot.

precision, it is more difficult to check the results from interlaboratory comparisons. Incidentally the second laboratory will also get good checks. Evidently each laboratory has good precision but something operates to displace the results in one laboratory from those obtained in another. It is appropriate to assume that each laboratory has its own systematic error.

A laboratory that has large systematic errors will be represented by a point x, y in the Youden plot that sticks to the 45° axis but that lies quite a distance from x_0, y_0. The distance of the point $T(x_1, y_1)$ (Figure 3.18) to the 45° axis is an estimate of the random error (the precision) of this laboratory. The distance from S (the projection of T on the 45° axis) to x_0, y_0 is an estimate of the systematic error.

In fact the Youden plot is a graphic representation of an analysis of variance (Section 4.6). The correlation between these methods can be seen by comparing some parameters.

In an analysis of variance the random error σ_0, is estimated by

$$\sigma_0^2 = \sum_{i=1}^{k} \sum_{j=1}^{n} \frac{(x_{ij} - \bar{x}_i)^2}{k(n-1)} = \frac{\sum_{i=1}^{k} \sum_{j=1}^{n} (x_{ij} - \bar{x}_i)^2}{k} \qquad (3.74)$$

where k = number of laboratories

n = number of analyses for each laboratory (here $n = 2$)

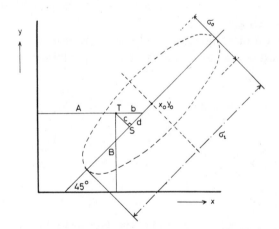

Figure 3.18. Precision and accuracy in the Youden plot.

In the Youden plot (Figure 3.18)

$$\sum_{i=1}^{k}\sum_{j=1}^{n}(x_{ij}-\bar{x}_i)^2 = \sum_{i=1}^{k}\{[A_i-(A_i+\tfrac{1}{2}b_i)]^2+[B_i-(B_i-\tfrac{1}{2}b_i)]^2\}$$

$$= \sum_{i=1}^{k}[(\tfrac{1}{2}b_i)^2+(\tfrac{1}{2}b_i)^2] \tag{3.75}$$

$$= \sum_{i=1}^{k}\tfrac{1}{2}b_i^2 = \sum_{i=1}^{k}c_i^2$$

$$\rightarrow \sigma_0^2 = \frac{\sum_{i=1}^{k}c_i^2}{k(n-1)} \tag{3.76}$$

In this case $n = 2$, so

$$\sigma_0^2 = \frac{\sum_{i=1}^{k}c_i^2}{k} \tag{3.77}$$

In the analysis of variance the total error $\sigma_0^2 + n\sigma_1^2$ is estimated by

$$\sigma_0^2 + n\sigma_1^2 = n\frac{\sum_{i=1}^{k}(\bar{x}_i-\bar{x})^2}{k-1} \tag{3.78}$$

With $n = 2$

$$\sigma_0^2 + 2\sigma_1^2 = 2\frac{\sum_{i=1}^{k}(\bar{x}_i-\bar{x})^2}{k-1} \tag{3.79}$$

In the Youden plot,

$$\sigma_0^2 + 2\sigma_1^2 = 2\frac{\sum_{i=1}^{k}d_i^2}{k-1} \tag{3.80}$$

The significance of the difference between total error and precision σ_0 can be tested by applying the F-test:

$$F = \frac{2\left(\sum\limits_{i=1}^{k} d_i^2\right) \Big/ (k-1)}{\left(\sum\limits_{i=1}^{k} c_i^2\right) \Big/ k} \tag{3.81}$$

For the value of F greater than the tabulated F [for a reliability α, $v_1 = k-1$ and $v_2 = k$ (Table 3.16)] it may be supposed that a systematic error exists:

$$\sigma_1^2 = \left[\frac{\sum\limits_{i=1}^{k} d_i^2}{k-1}\right] = \left[\frac{\frac{1}{2}\left(\sum\limits_{i=1}^{k} c_i^2\right)}{k}\right] \tag{3.82}$$

The Youden plot can be used not only for comparison of the quality of laboratories with regard to their results but also for comparison of the quality of two methods of analysis with regard to their (in)sensitivity to various laboratories. When one is comparing laboratories it is important that a sufficient number of them be involved, because only in such situations can quantitative results be obtained.

In practice, however, often only the qualitative result is presented in graphic form. It should be noted that this works quite well because "outliers," i.e., laboratories having points far from x_0, y_0 along the 45° axis, seldom appear at the same spot when the round-robin test is repeated.

It must be emphasized that quantitative results can be obtained only when the distribution of the data is not significantly different from a normal one. For a skewed distribution, as may be the case when the samples x and y are near the limit of detection, an appropriate linearizing method should be applied (Section 4.1.2) or the quantitative presentation should be abandoned in favor of the graphic plot. Remember that the method is not an absolute one. The inclusion of many deviating laboratories can give the impression that the "good" ones are the outliers. The use of standard samples can give a more reliable result. However, it must be emphasized that the success of the

Table 3.16. Probability Points of the Variance Ratio (F Distribution)

Probability Point		ϕ_N (corresponding to greater mean square)																			Probability Point	ϕ_D
ϕ_D		1	2	3	4	5	6	7	8	9	10	12	15	20	24	30	40	60	120	∞		
1	0.100	39.9	49.5	53.6	55.8	57.2	58.2	58.9	59.4	59.9	60.2	60.7	61.2	61.7	62.0	62.3	62.5	62.8	63.1	63.3	0.100	1
	0.050	161	199	216	225	230	234	237	239	241	242	244	246	248	249	250	251	252	253	254	0.050	
	0.025	648	800	864	900	922	937	948	957	963	969	977	985	993	997	1001	1006	1010	1014	1018	0.025	
	0.010	4052	4999	5403	5625	5764	5859	5928	5982	6022	6056	6106	6157	6209	6235	6261	6287	6313	6339	6366	0.010	
2	0.100	8.53	9.00	9.16	9.24	9.29	9.33	9.35	9.37	9.38	9.39	9.41	9.42	9.44	9.45	9.46	9.47	9.47	9.48	9.49	0.100	2
	0.050	18.5	19.0	19.2	19.2	19.3	19.3	19.4	19.4	19.4	19.4	19.4	19.4	19.4	19.5	19.5	19.5	19.5	19.5	19.5	0.050	
	0.025	38.5	39.0	39.2	39.2	39.3	39.3	39.4	39.4	39.4	39.4	39.4	39.4	39.4	39.5	39.5	39.5	39.5	39.5	39.5	0.025	
	0.010	98.5	99.0	99.2	99.2	99.3	99.3	99.4	99.4	99.4	99.4	99.4	99.4	99.4	99.5	99.5	99.5	99.5	99.5	99.5	0.010	
3	0.100	5.54	5.46	5.39	5.34	5.31	5.28	5.27	5.25	5.24	5.23	5.22	5.20	5.18	5.18	5.17	5.16	5.15	5.14	5.13	0.100	3
	0.050	10.1	9.55	9.28	9.12	9.01	8.94	8.89	8.85	8.81	8.79	8.74	8.70	8.66	8.64	8.62	8.59	8.57	8.55	8.53	0.050	
	0.025	17.4	16.0	15.4	15.1	14.9	14.7	14.6	14.5	14.5	14.4	14.3	14.3	14.2	14.1	14.1	14.0	14.0	13.9	13.9	0.025	
	0.010	34.1	30.8	29.5	28.7	28.2	27.9	27.7	27.5	27.3	27.2	27.1	26.9	26.7	26.6	26.5	26.4	26.3	26.2	26.1	0.010	
4	0.100	4.54	4.32	4.19	4.11	4.05	4.01	3.98	3.95	3.94	3.92	3.90	3.87	3.84	3.83	3.82	3.80	3.79	3.78	3.76	0.100	4
	0.050	7.71	6.94	6.59	6.39	6.26	6.16	6.09	6.04	6.00	5.96	5.91	5.86	5.80	5.77	5.75	5.72	5.69	5.66	5.63	0.050	
	0.025	12.2	10.6	10.0	9.60	9.36	9.20	9.07	8.98	8.90	8.84	8.75	8.66	8.56	8.51	8.46	8.41	8.36	8.31	8.26	0.025	
	0.010	21.2	18.0	16.7	16.0	15.5	15.2	15.0	14.8	14.7	14.5	14.4	14.2	14.0	13.9	13.8	13.7	13.7	13.6	13.5	0.010	
5	0.100	4.06	3.78	3.62	3.52	3.45	3.40	3.37	3.34	3.32	3.30	3.27	3.24	3.21	3.19	3.17	3.16	3.14	3.12	3.10	0.100	5
	0.050	6.61	5.79	5.41	5.19	5.05	4.95	4.88	4.82	4.77	4.74	4.68	4.62	4.56	4.53	4.50	4.46	4.43	4.40	4.36	0.050	
	0.025	10.0	8.43	7.76	7.39	7.15	6.98	6.85	6.76	6.68	6.62	6.52	6.43	6.33	6.28	6.23	6.18	6.12	6.07	6.02	0.025	
	0.010	16.3	13.3	12.1	11.4	11.0	10.7	10.5	10.3	10.2	10.1	9.89	9.72	9.55	9.47	9.38	9.29	9.20	9.11	9.02	0.010	
6	0.100	3.78	3.46	3.29	3.18	3.11	3.05	3.01	2.98	2.96	2.94	2.90	2.87	2.84	2.82	2.80	2.78	2.76	2.74	2.72	0.100	6
	0.050	5.99	5.14	4.76	4.53	4.39	4.28	4.21	4.15	4.10	4.06	4.00	3.94	3.87	3.84	3.81	3.77	3.74	3.70	3.67	0.050	
	0.025	8.81	7.26	6.60	6.23	5.99	5.82	5.70	5.60	5.52	5.46	5.37	5.27	5.17	5.12	5.07	5.01	4.96	4.90	4.85	0.025	
	0.010	13.7	10.9	9.78	9.15	8.75	8.47	8.26	8.10	7.98	7.87	7.72	7.56	7.40	7.31	7.23	7.14	7.06	6.97	6.88	0.010	
7	0.100	3.59	3.26	3.07	2.96	2.88	2.83	2.78	2.75	2.72	2.70	2.67	2.63	2.59	2.58	2.56	2.54	2.51	2.49	2.47	0.100	7
	0.050	5.59	4.74	4.35	4.12	3.97	3.87	3.79	3.73	3.68	3.64	3.57	3.51	3.44	3.41	3.38	3.34	3.30	3.27	3.23	0.050	
	0.025	8.07	6.54	5.89	5.52	5.29	5.12	4.99	4.90	4.82	4.76	4.67	4.57	4.47	4.42	4.36	4.31	4.25	4.20	4.14	0.025	
	0.010	12.2	9.55	8.45	7.85	7.46	7.19	6.99	6.84	6.72	6.62	6.47	6.31	6.16	6.07	5.99	5.91	5.82	5.74	5.65	0.010	

Table 3.16. (Continued)

Probability Point	ϕ_D	1	2	3	4	5	6	7	8	9	10	12	15	20	24	30	40	60	120	∞	ϕ_D	Probability Point
									ϕ_N (corresponding to greater mean square)													
0.100	8	3.46	3.11	2.92	2.81	2.73	2.67	2.62	2.59	2.56	2.54	2.50	2.46	2.42	2.40	2.38	2.36	2.34	2.32	2.29	8	0.100
0.050		5.32	4.46	4.07	3.94	3.69	3.58	3.50	3.44	3.39	3.35	3.28	3.22	3.15	3.12	3.08	3.04	3.01	2.97	2.93		0.050
0.025		7.57	6.06	5.42	5.05	4.82	4.65	4.53	4.43	4.36	4.30	4.20	4.10	4.00	3.95	3.89	3.84	3.78	3.73	3.67		0.025
0.010		11.3	8.65	7.59	7.01	6.63	6.37	6.18	6.03	5.91	5.81	5.67	5.52	5.36	5.28	5.20	5.12	5.03	4.95	4.86		0.010
0.100	9	3.36	3.01	2.81	2.69	2.61	2.55	2.51	2.47	2.44	2.42	2.38	2.34	2.30	2.28	2.25	2.23	2.21	2.18	2.16	9	0.100
0.050		5.12	4.26	3.86	3.63	3.48	3.37	3.29	3.23	3.18	3.14	3.07	3.01	2.94	2.90	2.86	2.83	2.79	2.75	2.71		0.050
0.025		7.21	5.71	5.08	4.72	4.48	4.32	4.20	4.10	4.03	3.96	3.87	3.77	3.67	3.61	3.56	3.51	3.45	3.39	3.33		0.025
0.010		10.6	8.02	6.99	6.42	6.06	5.80	5.61	5.47	5.35	5.26	5.11	4.96	4.81	4.73	4.65	4.57	4.48	4.40	4.31		0.010
0.100	10	3.28	2.92	2.73	2.61	2.52	2.46	2.41	2.38	2.35	2.32	2.28	2.24	2.20	2.18	2.16	2.13	2.11	2.08	2.06	10	0.100
0.050		4.96	4.10	3.71	3.48	3.33	3.22	3.14	3.07	3.02	2.98	2.91	2.84	2.77	2.74	2.70	2.66	2.62	2.58	2.54		0.050
0.025		6.94	5.46	4.83	4.47	4.24	4.07	3.95	3.85	3.78	3.72	3.62	3.52	3.42	3.37	3.31	3.26	3.20	3.14	3.08		0.025
0.010		10.0	7.56	6.55	5.99	5.64	5.39	5.20	5.06	4.94	4.85	4.71	4.56	4.41	4.33	4.25	4.17	4.08	4.00	3.91		0.010
0.100	12	3.18	2.81	2.61	2.48	2.39	2.33	2.28	2.24	2.21	2.19	2.15	2.10	2.06	2.04	2.01	1.99	1.96	1.93	1.90	12	0.100
0.050		4.75	3.89	3.49	3.26	3.11	3.00	2.91	2.85	2.80	2.75	2.69	2.62	2.54	2.51	2.47	2.43	2.38	2.34	2.30		0.050
0.025		6.55	5.10	4.47	4.12	3.89	3.73	3.61	3.51	3.44	3.37	3.28	3.18	3.07	3.02	2.96	2.91	2.85	2.79	2.72		0.025
0.010		9.33	6.93	5.95	5.41	5.06	4.82	4.64	4.50	4.39	4.30	4.16	4.01	3.86	3.78	3.70	3.62	3.54	3.45	3.36		0.010
0.100	15	3.07	2.70	2.49	2.36	2.27	2.21	2.16	2.12	2.09	2.06	2.02	1.97	1.92	1.90	1.87	1.85	1.82	1.79	1.76	15	0.100
0.050		4.54	3.68	3.29	3.06	2.90	2.79	2.71	2.64	2.59	2.54	2.48	2.40	2.33	2.29	2.25	2.20	2.16	2.11	2.07		0.050
0.025		6.20	4.77	4.15	3.80	3.58	3.41	3.29	3.20	3.12	3.06	2.96	2.86	2.76	2.70	2.64	2.59	2.52	2.46	2.40		0.025
0.010		8.68	6.36	5.42	4.89	4.56	4.32	4.14	4.00	3.89	3.80	3.67	3.52	3.37	3.29	3.21	3.13	3.05	2.96	2.87		0.010
0.100	20	2.97	2.59	2.38	2.25	2.16	2.09	2.04	2.00	1.96	1.94	1.89	1.84	1.79	1.77	1.74	1.71	1.68	1.64	1.61	20	0.100
0.050		4.35	3.49	3.10	2.87	2.71	2.60	2.51	2.45	2.39	2.35	2.28	2.20	2.12	2.08	2.04	1.99	1.95	1.90	1.84		0.050
0.025		5.87	4.46	3.86	3.51	3.29	3.13	3.01	2.91	2.84	2.77	2.68	2.57	2.46	2.41	2.35	2.29	2.22	2.16	2.09		0.025
0.010		8.10	5.85	4.94	4.43	4.10	3.87	3.70	3.56	3.46	3.37	3.23	3.09	2.94	2.86	2.78	2.69	2.61	2.52	2.42		0.010

138

ν_2	α	1	2	3	4	5	6	7	8	9	10	12	15	20	24	30	40	60	120	∞
24	0.100	2.93	2.54	2.33	2.19	2.10	2.04	1.98	1.94	1.91	1.88	1.83	1.78	1.73	1.70	1.67	1.64	1.61	1.57	1.53
	0.050	4.26	3.40	3.01	2.78	2.62	2.51	2.42	2.36	2.30	2.25	2.18	2.11	2.03	1.98	1.94	1.89	1.84	1.79	1.73
	0.025	5.72	4.32	3.72	3.38	3.15	2.99	2.87	2.78	2.70	2.64	2.54	2.44	2.33	2.27	2.21	2.15	2.08	2.01	1.94
	0.010	7.82	5.61	4.72	4.22	3.90	3.67	3.50	3.36	3.26	3.17	3.03	2.89	2.74	2.66	2.58	2.49	2.40	2.31	2.21
30	0.100	2.88	2.49	2.28	2.14	2.05	1.98	1.93	1.88	1.85	1.82	1.77	1.72	1.67	1.64	1.61	1.57	1.54	1.50	1.46
	0.050	4.17	3.32	2.92	2.69	2.53	2.42	2.33	2.27	2.21	2.16	2.09	2.01	1.93	1.89	1.84	1.79	1.74	1.68	1.62
	0.025	5.57	4.18	3.59	3.25	3.03	2.87	2.75	2.65	2.57	2.51	2.41	2.31	2.20	2.14	2.07	2.01	1.94	1.87	1.79
	0.010	7.56	5.39	4.51	4.02	3.70	3.47	3.30	3.17	3.07	2.98	2.84	2.70	2.55	2.47	2.39	2.30	2.21	2.11	2.01
40	0.100	2.84	2.44	2.23	2.09	2.00	1.93	1.87	1.83	1.79	1.76	1.71	1.66	1.61	1.57	1.54	1.51	1.47	1.42	1.38
	0.050	4.08	3.23	2.84	2.61	2.45	2.34	2.25	2.18	2.12	2.08	2.00	1.92	1.84	1.79	1.74	1.69	1.64	1.58	1.51
	0.025	5.42	4.05	3.46	3.13	2.90	2.74	2.62	2.53	2.45	2.39	2.29	2.18	2.07	2.01	1.94	1.88	1.80	1.72	1.64
	0.010	7.31	5.18	4.31	3.83	3.51	3.29	3.12	2.99	2.89	2.80	2.66	2.52	2.37	2.29	2.20	2.11	2.02	1.92	1.80
60	0.100	2.79	2.39	2.18	2.04	1.95	1.87	1.82	1.77	1.74	1.71	1.66	1.60	1.54	1.51	1.48	1.44	1.40	1.35	1.29
	0.050	4.00	3.15	2.76	2.53	2.37	2.25	2.17	2.10	2.04	1.99	1.92	1.84	1.75	1.70	1.65	1.59	1.53	1.47	1.39
	0.025	5.29	3.93	3.34	3.01	2.79	2.63	2.51	2.41	2.33	2.27	2.17	2.06	1.94	1.88	1.82	1.74	1.67	1.58	1.48
	0.010	7.08	4.98	4.13	3.65	3.34	3.12	2.95	2.82	2.72	2.63	2.50	2.35	2.20	2.12	2.03	1.94	1.84	1.73	1.60
120	0.100	2.75	2.35	2.13	1.99	1.90	1.82	1.77	1.72	1.68	1.65	1.60	1.55	1.48	1.45	1.41	1.37	1.32	1.26	1.19
	0.050	3.92	3.07	2.68	2.45	2.29	2.18	2.09	2.02	1.96	1.91	1.83	1.75	1.66	1.61	1.55	1.50	1.43	1.35	1.25
	0.025	5.15	3.80	3.23	2.89	2.67	2.52	2.39	2.30	2.22	2.16	2.05	1.94	1.82	1.76	1.69	1.61	1.53	1.43	1.31
	0.010	6.85	4.79	3.95	3.48	3.17	2.96	2.79	2.66	2.56	2.47	2.34	2.19	2.03	1.95	1.86	1.76	1.66	1.53	1.38
∞	0.100	2.71	2.30	2.08	1.94	1.85	1.77	1.72	1.67	1.63	1.60	1.55	1.49	1.42	1.38	1.34	1.30	1.24	1.17	1.00
	0.050	3.84	3.00	2.60	2.37	2.21	2.10	2.01	1.94	1.88	1.83	1.75	1.67	1.57	1.52	1.46	1.39	1.32	1.22	1.00
	0.025	5.02	3.69	3.12	2.79	2.57	2.41	2.29	2.19	2.11	2.05	1.94	1.83	1.71	1.64	1.57	1.48	1.39	1.27	1.00
	0.010	6.63	4.61	3.78	3.32	3.02	2.80	2.64	2.51	2.41	2.32	2.18	2.04	1.88	1.79	1.70	1.59	1.47	1.32	1.00

139

method does not depend on marking the absolute outliers. The effect is in the first place psychological: participants (indicated only by a code) want to improve their quality; they do not want to be punished or made ashamed. Table 3.17 gives an abstract of the IUPAC-1987 protocol for collaborative studies that does not use the graphic output.

Table 3.17. IUPAC-1987 Harmonized Protocol for the Design, Conduct, and Interpretation of Collaborative Studies[a]

1. The results of collaborative studies should be analyzed by one-way analysis of variance ("material by material"), but more complex analyses are not precluded.

2. The absolute minimum number of materials to be used in a collaborative study is five. However, when a single level specification is involved this may be reduced to an absolute minimum of three.

3. The most important objective of a collaborative study should be attaining a reliable estimate of reproducibility[b] parameters. To the extent that the performance of known (parallel) replicate analyses detracts from this objective, a requirement for the use of this type of replicate analysis should be discouraged.

4. The best estimate of repeatability[b] parameters is obtained by the following procedures (listed approximately in order of desirability):
   a. The use of a split-level design (single values from the analysis of each of two closely related materials)
   b. The use of both split levels and blind duplicate analyses in the same study
   c. The use of blind duplicate analyses
   d. The use of known duplicate analyses (two repeat analyses from the same test sample), but only when it is not practical to use one of the preceding designs

5. The minimum number of participating laboratories in a collaborative study is eight. Only when it is impossible to obtain this number may the study be conducted with fewer, but with an absolute minimum of five laboratories. (Although it is desirable to have more than eight laboratories participating, studies containing more than about fifteen laboratories become unwieldy.)

6. The precision estimates are to be calculated both with no outliers removed and with outliers removed, using the Cochran and Grubbs outlier tests. The Grubbs tests should be applied only to laboratory means, not to individual values of replicated designs. Outlier removal should be stopped when more than 22% (i.e., more than two out of nine laboratories) would be removed as a result of the sequential application of the outlier tests.

7. When standard deviations are expressed in relation to the absolute value of the arithmetic mean, the term recommended is relative standard deviation (RSD). The equivalent ISO term is coefficient of variation (CV).

[a] Only the main requirements outlined in the document are given below.
[b] As defined in ISO 5725-1986.

3.4.4. The Ranking Test

Another way to assess the quality of a laboratory or an analytical method is the application of the ranking test (DA 84, YO 72). The data from a collaborative test may be arranged in a classification scheme as shown in Table 3.18. The right-hand side of the table shows the data substituted by rankings assigned to the laboratories according to the measured concentration for a series of samples. Rank 1 is given to the largest amount, rank 2 to the next largest, and so on. When two laboratories have the same score for the xth place, each laboratory is assigned the rank $x + \frac{1}{2}$. For a triple tie for the xth place, all three laboratories get the rank $x + 1$. The laboratory's score equals the sum of the ranks it receives.

$$M = \sum_{i=1}^{n} R_i \qquad (3.83)$$

where R_i = ranking for sample i

n = number of quality parameters

Table 3.18. Ranking Test

Coll. No.	Results (%) Samples					Ranked Results Samples					Coll. Score
	1	2	3	4	5	1	2	3	4	5	
7	4.59	1.46	5.64	2.19	27.32	9	5.5	6	4	3	27.5
8	4.94	1.52	5.68	2.28	26.44	1	1	3	2	10	17
9	4.80	1.40	5.62	2.12	26.89	3.5	8.5	7.5	6.5	8	34
10	4.73	1.46	5.65	2.09	27.17	5	5.5	5	8	4	27.5
11	4.72	1.51	5.62	2.12	27.00	6.5	2.5	7.5	6.5	6	29
12	4.80	1.51	5.80	2.29	27.48	3.5	2.5	1	1	1	9[a]
13	4.45	1.40	5.45	2.07	27.02	10	8.5	10	9	5	42.5
15	4.72	1.50	5.58	2.27	26.76	6.5	4	9	3	9	31.5
16	4.63	1.32	5.69	2.04	26.92	8	10	2	10	7	37
17	4.88	1.42	5.67	2.16	27.39	2	7	4	5	2	20

Source: Reprinted with permission from W. J. Youden and E. H. Steiner, *Statistical Manual of the Association of Official Analytical Chemists, AOAC*, Washington, DC, 1975.

[a] Designates unusually low score.

For n samples, the minimum score is n and the maximum score is nm, where m denotes the number of participating laboratories. A laboratory that scores the highest amount on every one of the n materials gets the score of n. Such a score is obviously associated with a laboratory that consistently gets high results, and the presumption is that this laboratory has a pronounced systematic error. When the results of the laboratories stem from a random process, the scores tend to cluster

Table 3.19. Approximate 5% Two-Tail Limits for Ranking Scores

No. of Labs.	No. of Materials												
	3	4	5	6	7	8	9	10	11	12	13	14	15
3	—	4	5	7	8	10	12	13	15	17	19	20	22
	—	12	15	17	20	22	24	27	29	31	33	36	38
4	—	4	6	8	10	12	14	16	18	20	22	24	26
	—	16	19	22	25	28	31	34	37	40	43	46	49
5	—	5	7	9	11	13	16	18	21	23	26	28	31
	—	19	23	27	31	35	38	42	45	49	52	56	59
6	3	5	7	10	12	15	18	21	23	26	29	32	35
	18	23	28	32	37	41	45	49	54	58	62	66	70
7	3	5	8	11	14	17	20	23	26	29	32	36	39
	21	27	32	37	42	47	52	57	62	67	72	76	81
8	3	6	9	12	15	18	22	25	29	32	36	39	43
	24	30	36	42	48	54	59	65	70	76	81	87	92
9	3	6	9	13	16	20	24	27	31	35	39	43	47
	27	34	41	47	54	60	66	73	79	85	91	97	103
10	4	7	10	14	17	21	26	30	34	38	43	47	51
	29	37	45	52	60	67	73	80	87	94	100	107	114
11	4	7	11	15	19	23	27	32	36	41	46	51	55
	32	41	49	57	65	73	81	88	96	103	110	117	125
12	4	7	11	15	20	24	29	34	39	44	49	54	59
	35	45	54	63	71	80	88	96	104	112	120	128	136
13	4	8	12	16	21	26	31	36	42	47	52	58	63
	38	48	58	68	77	86	95	104	112	121	130	138	147
14	4	8	12	17	22	27	33	38	44	50	56	61	67
	41	52	63	73	83	93	102	112	121	130	139	149	158
15	4	8	13	18	23	29	35	41	47	53	59	65	71
	44	56	67	78	89	99	109	119	129	139	149	159	169

Source: Reprinted with permission from W. J. Youden and E. H. Steiner, *Statistical Manual of the Association of Official Analytical Chemists, AOAC*, Washington, DC, 1975.

around the value $n(m + 1)/2$. The statistical distribution of such scores has been tabulated in Table 3.19 (YO 72).

The upper and lower limits shown in this table have been determined to be based on the assumption that there is a 95% reliability that the rankings between these limits result from a random process. In extreme situations most of the laboratories may get approximately the same ranking on each material. Here the scores approximate the values $n, 2n, \ldots, mn$. Obviously this is an indictment of the analytical method: presumably either its description is inadequately written or its implication is unacceptably sensitive to the environmental differences encountered in the various laboratories.

3.5. SEQUENTIAL ANALYSIS

When performing a series of experiments, one sometimes has to decide whether the experiment gives a result that confirms the hypothesis that is tested or whether, alternatively, this hypothesis must be rejected. This testing can be done by applying statistics to the set of observations. There are then three possible decisions:

1. From the results obtained it is not yet possible to reject the hypothesis nor to accept it.
2. The hypothesis is rejected.
3. The hypothesis is accepted.

With the use of the proper statistics, decisions 2 and 3 can be made with a certain degree of confidence. When the experiment has been planned without prior knowledge of the statistical behavior of the experiment, it is possible that decisions 2 and 3 will be made with a confidence level that is much higher than required for the problem under investigation. The conclusion should be that the number of observations could have been less. If, however, decision 1 is the result, this means that the experiment has to be continued. The question to be answered is now: how long should the experiment be continued in order to reach decisions 2 or 3 with sufficient—but not too much—confidence?

It is obvious that the design of an experiment consisting of a series of observations is more critical when the cost—in money or time—per observation is high.

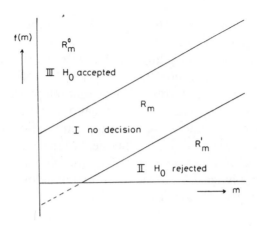

Figure 3.19. Sequential analysis, decision areas.

A method that circumvents the problems just described is sequential analysis. By this method a test is conducted that allows a decision to be made after each observation. Thus the number of observations is minimal. The theory of this method is described in many textbooks (e.g., DA 84, WA 47). Application to various problems requires a thorough knowledge of statistics. Here we describe only two cases, problems that are often encountered in analytical chemistry.

Sequential analysis is a method used to guide statistical interference in observation series in order to limit the number of observations of that series, which is not known in advance. The decision as to whether to terminate the experiment depends on the result of the observations collected so far. A comparison of sequential and nonsequential analysis is given in Table 3.20. In a sequential test the number of observations required depends on the result of the observations and therefore is not fixed in advance. After each observation a decision is made whether the null hypothesis H_0 will be accepted or rejected or if the experiment will be continued. The decision to terminate the experiment at any moment depends on the observations made so far.

3.5.1. Mathematical Description

Suppose a hypothesis H has to be tested by performing an experiment that results in the variable x. Then $x_1, x_2, \ldots, x_m$ are m observations of the variable x.

Table 3.20. Comparison of Sequential and Nonsequential Analyses

Sequential Analysis	Nonsequential Analysis
The number of observations required is not known in advance.	The number of observations is known before the experiment starts.
After each observation one out of three possible decisions is selected: 1. Continue the experiment and make a new observation. 2. Reject the null hypothesis and terminate the experiment. 3. Accept the null hypothesis and terminate the experiment.	After the experiment, comprising a series of observations, three decisions are possible: 1. Start a new series of observations. 2. Reject the null hypothesis. 3. Accept the null hypothesis.
After each observation the decision to terminate the experiment may result.	The experiment is terminated at the end of the series of observations.
The decision to terminate the experiment as a rule results in fewer observations than in nonsequential analysis.	
The method is recommended when the measuring time is short compared with computing time.	The method is recommended when the measuring time is long compared with computing time.
The method is preferred for expensive measurements.	

The progress of the test can be shown graphically by computing the value of a function of all observations, $F(m)$. This value is plotted as a function of the number of observations. The sample space M_m can be subdivided into three mutually exclusive subspaces:

R_m°: space where H is accepted

R_m': space where H is rejected

R_m: space where no decision is possible

In Figure 3.19 two boundaries are indicated. The direction of these boundaries depends on the values of α and β and the parameters tested (α = probability of an error of the first kind—reject H wrongly; β = probability of an error of the second kind—accept H wrongly). When the function value crosses one of the boundaries and enters zone

II or III, the experiment is terminated by the decision dictated by the zone number.

Example. Suppose a random variable x with a probability distribution $f(x)$ depends on one unknown parameter θ: $\{f(x, \theta)\}$. The null hypothesis reads H_0: $\theta = \theta_0$. The alternative is H_1: $\theta = \theta_1$. If H_0 is accepted, the distribution of x is $f(x, \theta_0)$; if H_1 is accepted it is $f(x, \theta_1)$. With every positive integer m the probability that a series of observations $x_1, x_2, \ldots, x_m$ results is

$$P_{0,m} = f(x_1, \theta_0), \ldots, f(x_m, \theta_0) \quad \text{when } H_0 \text{ is accepted}$$
$$P_{1,m} = f(x_1, \theta_0), \ldots, f(x_m, \theta_0) \quad \text{when } H_1 \text{ is accepted}$$

To test whether the series of observations has to be continued the ratio $P_{1,m}/P_{0,m}$ is computed after each measurement and compared with two constants A and B. These constants are defined as

$$A = (1 - \beta)/\alpha \quad \text{and} \quad B = \beta/(1 - \alpha)$$

Now the three possible situations are as follows:

$$\textbf{I.} \quad B < (P_{1,m}/P_{0,m}) < A \tag{3.84}$$

The series of observations is extended with an observation x_{m+1}.

$$\textbf{II.} \quad (P_{1,m}/P_{0,m}) \geqslant A \tag{3.85}$$

The experiment is terminated with rejection of H_0 and acceptance of H_1.

$$\textbf{III.} \quad (P_{1,m}/P_{0,m}) \leqslant B \tag{3.86}$$

The experiment is terminated with accpetance of H_0 and rejection of H_1.

Often eq. 3.84 is logarithmized thus:

$$\log B < \log(P_{1,m}/P_{0,m}) < \log A \tag{3.87}$$

in which

$$\log\left(\frac{P_{1,m}}{P_{0,m}}\right) = \log\frac{f(x_1\theta_1),\ldots,f(x_m\theta_1)}{f(x_1\theta_0),\ldots,f(x_m\theta_0)}$$

$$= \log\frac{f(x_1\theta_1)}{f(x_1\theta_0)} + \log\frac{f(x_2\theta_1)}{f(x_2\theta_0)} + \cdots + \log\frac{f(x_m\theta_1)}{f(x_m\theta_0)} \tag{3.88}$$

Substituting z_i according to

$$z_i = \log\{f(x_i\theta_1)/f(x_i\theta_0)\} \tag{3.89}$$

results in

$$\log B < \sum_{i=1}^{m} z_i < \log A \tag{3.90}$$

An example is the test for accuracy, that is, whether a method of analysis has the same mean as a standard method or is higher or lower. A normal distribution with standard deviation σ for both methods is assumed, the mean of the standard method being x_0. A bias Δ must be detected with given confidence α and β. This is a combination of two tests: a test on a deviation $> +\Delta$ and a test on a deviation $> |-\Delta|$. Therefore the value of α, the probability that the null hypothesis will be rejected, will be halved; thus $\alpha' = \alpha/2$.

A. The first test is on a positive deviation of more than Δ. The null hypothesis is

$$H_0:\quad x = x_0 = \mu$$
$$H_1:\quad x_1 = x_0 + \Delta$$

The experiment is continued when $x_0 < x < x_1$.

Since we assumed a Gaussian distribution, the probability that a series of m observations leads to acceptance of H_0 reads

$$P_{0,m} = k\exp\left[-\frac{1}{2\sigma^2}\sum_{q=1}^{m}(x_q - x_0)^2\right] \tag{3.91}$$

where σ = the standard deviation
k = constant

The probability that H_0 is rejected is

$$P_{1,m} = k \exp\left[-\frac{1}{2\sigma^2} \sum_{q=1}^{m} (x_q - x_1)^2 \right] \tag{3.92}$$

and

$$\frac{P_{1,m}}{P_{0,m}} = \frac{\exp\left[-\left(\frac{1}{2}\sigma^2\right) \sum_{q=1}^{m} (x_q - x_1)^2 \right]}{\exp\left[-\left(\frac{1}{2}\sigma^2\right) \sum_{q=1}^{m} (x_q - x_0)^2 \right]}$$

$$= \exp \frac{1}{2\sigma^2}\left[-\sum_{q=1}^{m} (x_q - x_1)^2 + \sum_{q=1}^{m} (x_q - x_0)^2 \right] \tag{3.93}$$

$$= \frac{1}{2\sigma^2}\left[2(x_1 - x_0) \sum_{q=1}^{m} x_q - m(x_1^2 - x_0^2) \right]$$

$$= \exp \frac{\Delta}{2\sigma^2}\left[2 \sum_{q=1}^{m} x_q - m(2x_0 + \Delta) \right]$$

where $\Delta = x_1 - x_0$.

$$\ln \frac{P_{1,m}}{P_{0,m}} = \frac{\Delta}{2\sigma^2}\left[2 \sum_{q=1}^{m} x_q - m(2x_0 + \Delta) \right]$$

$$= \frac{\Delta}{\sigma^2}\left[\sum_{q=1}^{m} x_q - m(x_0 + \Delta/2) \right] \tag{3.94}$$

AI. The experiment is continued if (by eq. 3.94)

$$\ln B < \frac{\Delta}{\sigma^2} \sum_{q=1}^{m} x_q - \frac{m\Delta}{\sigma^2}(x_0 + \Delta/2) < \ln A \tag{3.95}$$

or

$$\frac{\sigma^2}{\Delta} \ln B + m(x_0 + \Delta/2) < \sum_{q=1}^{m} x_q < \frac{\sigma^2}{\Delta} \ln A + m(x_0 + \Delta/2) \tag{3.96}$$

AII. The experiment is terminated and H_0 rejected if

$$\sum_{q=1}^{m} x_q \geq \frac{\sigma^2}{\Delta} \ln A + m(x_0 + \Delta/2) \tag{3.97}$$

AIII. The experiment is terminated and H_0 accepted if

$$\sum_{q=1}^{m} x_q \leq \frac{\sigma^2}{\Delta} \ln B + m(x_0 + \Delta/2) \tag{3.98}$$

B. Analogous reasoning for the negative deviation leads to

$$H_0: \quad x = x_0$$
$$H_1: \quad x = x_1 = x_0 - \Delta$$

BI. The experiment is continued if

$$-\frac{\sigma^2}{\Delta} \ln A + m(x_0 - \Delta/2) < \sum_{q=1}^{m} x_q < -\frac{\sigma^2}{\Delta} \ln B + m(x_0 - \Delta/2) \tag{3.99}$$

BII. The experiment is terminated and H_0 rejected if

$$\sum_{q=1}^{m} x_q \leq -\frac{\sigma^2}{\Delta} \ln A + m(x_0 - \Delta/2) \tag{3.100}$$

BIII. The experiment is terminated and H_0 accepted if

$$\sum_{q=1}^{m} x_q \geq -\frac{\sigma^2}{\Delta} \ln B + m(x_0 - \Delta/2) \tag{3.101}$$

Combination of situations A and B can be visualized by plotting $\sum x_q$ as a function of m (Figure 3.20). Both decisions AI and BI lead to a continuation of the experiment. The boundaries L_0, L_1, L_0', and L_1' are given by, respectively,

$$L_0 = \frac{\sigma^2}{\Delta} \ln B + m(x_0 + \Delta/2) \quad \text{between decisions AI/AIII} \tag{3.102}$$

$$L_1 = \frac{\sigma^2}{\Delta} \ln A + m(x_0 + \Delta/2) \quad \text{between decisions AI/AII} \tag{3.103}$$

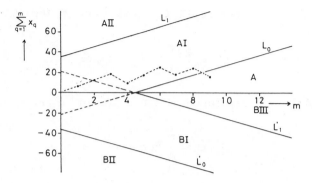

Figure 3.20. Sequential analysis of the accuracy of a series of observations (DA 70).

$$L_0' = \frac{\sigma^2}{\Delta} \ln A + m(x_0 - \Delta/2) \quad \text{between decisions BI/BII} \quad (3.104)$$

$$L_1' = \frac{\sigma^2}{\Delta} \ln B + m(x_0 - \Delta/2) \quad \text{between decisions BI/BIII} \quad (3.105)$$

An experiment was conducted in which the null hypothesis stated that the mean of the method should be equal to the mean of the standard method. A bias of $\Delta > 10$ should be detected with $\alpha = 0.05$ and $\beta = 0.10$. The mean of the standard method $x_0 = 0$, with $\sigma = 10$. Now

$$\ln A = \ln \frac{1 - \beta}{\alpha} = \ln \frac{0.90}{0.025} = 3.58$$

$$(3.106)$$

$$\ln B = \ln \frac{\beta}{1 - \alpha} = \ln \frac{0.10}{0.9755} = -2.28$$

This leads to the boundaries (Figure 3.20):

$$L_0 = -22.8 + 5m$$

$$L_1 = 35.8 + 5m$$

$$L_0' = -35.8 - 5m$$

$$L_1' = 22.8 - 5m$$

Table 3.21. Progress of a Sequential Analysis of Accuracy

No. of Observation (m)	Observation (x_q)	Sum $\left(\sum_{q=1}^{m} x_q\right)$	Decision	Conclusion[a] H_0	H_0'	Result
1	7	7	AI, BI	—	—	Continue
2	5	12	AI, BI	—	—	Continue
3	8	20	AI, BIII	—	a	Continue
4	−11	9	AI	—		Continue
5	10	19	AI	—		Continue
6	8	27	AI	—		Continue
7	−9	18	AI	—		Continue
8	6	24	AI	—		Continue
9	−7	17	AIII	a		Stop

[a] **a** = accepted; — = neither accepted nor rejected. No significant bias detected.

A series of observations as given in Table 3.21 can then be plotted (Figure 3.20)

C. A test on the precision of a method can also be conducted by sequential analysis. Now the null hypothesis is that the standard deviation σ of the method is equal to the standard deviation of a standard method, σ_0. The errors that are probable in the test are again α and β. The null hypothesis is rejected if $\sigma = \sigma_1 = \sigma_0 + \delta$.

$$H_0: \quad \sigma = \sigma_0$$
$$H_1: \quad \sigma = \sigma_1 = \sigma_0 + \delta$$

The probability that a series of observations $(x_1, x_2, \ldots, x_m)$ leads to acceptance of H_0 is

$$P_{0,m} = \frac{1}{(2\pi)^{m/2} \sigma_0^m} \exp\left[-\frac{1}{2\sigma_0^2} \sum_{q=1}^{m} (x_q - \mu)^2 \right] \qquad (3.107)$$

The probability of rejection of H_0 is

$$P_{1,m} = \frac{1}{(2\pi)^{m/2} \sigma_1^m} \exp\left[-\frac{1}{2\sigma_1^2} \sum_{q=1}^{m} (x_q - \mu)^2 \right] \qquad (3.108)$$

Now

$$\ln \frac{P_{1,m}}{P_{0,m}} = m \ln \frac{\sigma_0}{\sigma_1} + \left(\frac{1}{2\sigma_0^2} - \frac{1}{2\sigma_1^2} \right) \sum_{q=1}^{m} (x_q - \mu)^2 \qquad (3.109)$$

CI. The experiment is continued if

$$\ln B < \ln \frac{P_{1,m}}{P_{0,m}} < \ln A \tag{3.110}$$

$$\frac{2\ln B + m\ln(\sigma_1^2/\sigma_0^2)}{(1/\sigma_0^2) - (1/\sigma_1^2)} < \sum_{q=1}^{m} (x_q - \mu)^2 < \frac{2\ln A + m\ln(\sigma_1^2/\sigma_0^2)}{(1/\sigma_0^2) - (1/\sigma_1^2)} \tag{3.111}$$

or with

$$a_m = \frac{2\ln A + m\ln(\sigma_1^2/\sigma_0^2)}{(1/\sigma_0^2) - (1/\sigma_1^2)} \quad \text{and} \quad b_m = \frac{2\ln B + m\ln(\sigma_1^2/\sigma_0^2)}{(1/\sigma_0^2) - (1/\sigma_1^2)} \tag{3.112}$$

$$b_m < \sum_{q=1}^{m} (x_q - \mu)^2 < a_m \tag{3.113}$$

CII. The experiment is terminated and H_0 rejected if

$$\sum_{q=1}^{m} (x_q - \mu)^2 \geq a_m \tag{3.114}$$

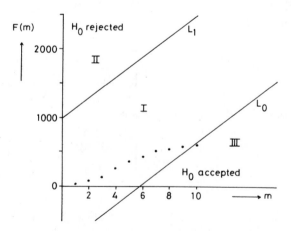

Figure 3.21. Sequential analysis of the precision of a series of observations.

Table 3.22. Progress of a Sequential Analysis of Precision

$\alpha = 0.05 \qquad \beta = 0.10 \qquad A = (1 - \beta)/\alpha = 0.90/0.05 = 18 \qquad B = \beta/(1 - \alpha)$

$= 0.10/0.95 = 0.015$

$\sigma_0 = 10 \qquad \sigma_1 = 15\mu = 0 \qquad L_0 = b_m \qquad L_1 = a_m$

$$L_0 = \frac{2\ln 0.105 + m\ln 2.25}{0.0055} = -811 + 146\,m$$

$$L_1 = \frac{2\ln 18 + m\ln 2.25}{0.0055} = 1040 + 146\,m$$

No. of Observation (m)	Observation (x_q)	Sum $\left[\displaystyle\sum_{q=1}^{m}(x_q - \mu)^2\right]$	Decision	Conclusion	Result
1	7	49	CI	—	Continue
2	5	74	CI	—	Continue
3	8	138	CI	—	Continue
4	−11	259	CI	—	Continue
5	10	359	CI	—	Continue
6	8	423	CI	—	Continue
7	−9	504	CI	—	Continue
8	6	540	CI	—	Continue
9	−7	589	CI	—	Continue
10	6	625	CIII	H_0 accepted	Stop

Table 3.23. Combined Sequential Analysis of Accuracy and Precision

No. of Observation (n, m)	Observation	Decision	Conclusion	Result
1	7	AI, BI, CI	—	Continue
2	5	AI, BI, CI	—	Continue
3	8	AI, BIII, CI	—	Continue
4	−11	AI, CI	—	Continue
5	10	AI, CI	—	Continue
6	8	AI, CI	—	Continue
7	−9	AI, CI	—	Continue
8	6	AI, CI	—	Continue
9	−7	AIII, CI	—	Continue
10	6	CIII	$(H_0)_A$, $(H_0)_B$, and $(H_0)_C$ are true	Stop

CIII. The experiment is terminated and H_0 accepted if

$$\sum_{q=1}^{m} (x_q - \mu)^2 \leqslant b_m \tag{3.115}$$

If the values of Table 3.22 are used, Figure 3.21 can be plotted.

The theory treated in the foregoing discussion shows that it is possible to test both accuracy and precision of an analytical method in one experiment.

However, the accuracy test is applicable only when the precision test has shown that the precision of the tested method and standard method is equal. Also it is possible to prove that the precision of both methods is equal only when the accuracy test has shown that no bias exists. If the sequential analysis scheme of the preceding chapter is used, this means that a condition for acceptance of the hypothesis that both precision and accuracy of the tested and standard method are equal is that AIII and BIII and CIII decisions should be made. This is depicted in Table 3.23, a combination of Table 3.21 and Table 3.22.

DATA PROCESSING

The first task of an analyst who wants to analyze a process or procedure is to collect relevant data of measurements on that process or procedure. The questions as to what data are relevant and how many measurements are required must be assumed from consideration of information theory and the sampling strategy (see Chapter 2 and Section 4.2). The reduction of the data collected according to a certain sampling strategy without loss of relevant information is the subject of this chapter. However, before the discussion of data reduction methods we must consider the methods of data production.

4.1. DATA PRODUCTION

There are various methods of data production:

1. A measurement is made and the outcome of the measurement is found from the position of a dial on a scale of an instrument. It is the analyst who estimates the position of the dial and converts this to numbers in a digital notation.
2. The measurement is made continuously and translated in the position of a recorder pen on a strip of recorder paper. The analyst reads the position of the recorded pattern on the time base at times of interest and converts the selected pen positions to digital numbers.
3. The measurement is made with a fixed frequency, converted electronically into a digital notation, and recorded on a device such as a disk or magnetic tape. The analyst is not involved in this process, apart from initiating the procedure of data collection.

The second task, after the numbers are obtained, is to reduce the number of measurements, remove the erroneous and irrelevant data, and convert the measurements into statements pertaining to the condition of the process or procedure under control. In modern analytical procedures, a data acquisition rate of 10^5 measurements per second is not unusual. In order to convert this enormous amount of data, collected, for example, during 1 day, into a form that can be handled, data reduction procedures should be applied. Depending on the aim of the analysis, various data reduction procedures may be applied.

4.1.1.   Tests for Normality

Since many statistical tests can be applied only when the set of data is normally (Gaussian) distributed, it is recommended that a test for normality be made first. Nowadays computations of statistical parameters are often performed routinely by application of standard calculator or computer subroutines. Nowhere in the instruction manuals is it adequately stressed that the application of these methods presupposes normal distribution.

In the literature several methods to test for normality have been described. An often used test is that for skewness. It signals asymmetry of the distribution. Although a passed test for skewness does not necessarily prove that the distribution is normal, a failed test obviously indicates that normal statistics should be applied with care.

A parameter for skewness is the so-called third moment, denoted by μ_3:

$$\mu_3 = \frac{\sum_{i=1}^{k} (x_i - \mu)^3}{n} \tag{4.1}$$

An index of skewness that is independent of the unit of measurement is the third moment divided by the cube of the standard deviation. This parameter is denoted by $\sqrt{\beta_1}$:

$$\sqrt{\beta_1} = \left(\frac{\mu_3}{\sigma^3}\right)^{1/2} \tag{4.2}$$

The value of $\sqrt{\beta_1}$ should be zero, but since σ normally is not known and is estimated by s, and because the third moment is also an estimation, some deviation from zero is allowed. Table 4.1 gives the probability of

Table 4.1. $\sqrt{\beta_1}$ **Values for Different Levels of Significance** α^a

n	$\alpha = .95$	$\alpha = .99$
25	0.711	1.061
30	0.661	0.982
35	0.621	0.921
40	0.587	0.869
45	0.558	0.825
50	0.533	0.787
60	0.492	0.723
70	0.459	0.673
80	0.432	0.631
90	0.409	0.596
100	0.389	0.567
150	0.321	0.464
200	0.280	0.403
300	0.230	0.329
400	0.200	0.285
500	0.179	0.255
1000	0.127	0.180

[a] Because the distribution of $\sqrt{\beta_1}$ is symmetrical about zero, the same values, with negative sign, correspond to the lower limits.

various deviations from zero for a normal distribution. Other, more reliable tests on normality are summarized in Table 4.2 (AM 77). As seen from this table, all methods are based on cumulative distribution functions, except the Shapiro–Wilk method. (Cumulative frequencies are obtained by adding for each class all the class frequencies up to that point, divided by the total number of data in the data set.) In the literature surveyed, there is no overall agreement as to which method is "best" for all possible empirical distributions. Comparison of the various normality tests results in the observation that the data set is considered to be normal by some methods and not normal by others. However, the chi-square test is generally regarded to be inferior to all other tests. Here only the Kolmogorov test is considered in more detail because it is generally applicable, easy to use, and one of the dependable methods.

Graphic Kolmogorov Test

1. Arrange the data in ascending order.
2. Draw the cumulative frequency line for a normal distribution (G in Figure 4.1).

Table 4.2. Principal Methods for Normality Testing

Test Method	Computing Formula	Symbols Used	Remarks	Refs.
Chi-square	$\chi^2 = \sum \left\{ \dfrac{(O_j - T_j)^2}{T_j} \right\}$	O_j = Observed class frequency T_j = Theoretical class frequency	The data are arranged in descending order and then subdivided into arbitrary classes, each containing at least 5 numbers. The "observed frequencies" are the number of data in each class. The "theoretical frequencies" are those based on the cumulative normal distribution function. The computed χ^2 is evaluated using a χ^2 table at $k - 3$ degrees of freedom, where k = number of classes. This method requires a relatively large sample size (at least 15–30) and is considered a test of low power.	DI 69, LI 67, CO 71
Lilliefors	$\max D_i^- = S(x_i) - (i-1)/n$	i = sequential data points or frequencies $(i-1)/n$ = cumulative distribution values of the data set $S(x_i)$ = corresponding normal cumulative distribution values	The data are arranged in ascending order and the D_j^- values computed by the formula given, i.e., not considering the first data cell. The largest D^{-j} value obtained is evaluated using the Lilliefors table. This method can be used with data sets given in terms of frequencies or individual data. Sample sizes as low as 4 can be tested. This method will fail if the first data cell contains a disproportionately large number of data compared to the rest of the set.	LI 67, CO 71
Kolmogorov D^+ (Kolmogorov one-sample statistics)	$\max D_i^+ = S(x_i) - i/n$	Symbols as given, and i/n = cumulative distribution values of the data set	The data are arranged in ascending order and the D_i^+ values computed, as given by the formula. The largest D_i^+ value obtained is evaluated using the	ST 70, ST 74, CO 71, MA 51,

				OW 62, SI 56

Kolmogorov table (two-sided). The method is usable with sample sizes as low as 2, with data given individually or as frequency distributions. This method can supplement the Lilliefors test if it is suspected that it may fail. A modified version of this test multiplies the largest D_i^+-value by a correction factor and evaluates this modified value using a special table.

| Kolmogorov D (modified Kolmogorov, E_n) | $D_i = \max(D^+, D^-)$ | Symbols as given | The largest D_i^+ or D_i^- calculated by the above formulas, is chosen. This value is multiplied by a correction factor and evaluated using a special table. Usable for individual data or those given as frequency distributions. | ST 70, ST 74 |

| Kuiper | $V = D^+ + D^-$ | Symbols as given | The sum of the largest D_i^+ and D_i^- is calculated by the formulas given. This sum is multiplied by a correction factor and evaluated using a special table. Usable for individual data or those given as frequency distributions. | ST 70, ST 74 |

| Cramer–von Mises (Cramer–Smirnov) | $W^2 = 1/(12n) + \sum [S(x_i) - (i-1)/n]^2$ | Symbols as given, and $(i-1)/n$ = cumulative distribution values of the data set | The data are arranged in ascending order. The W^2 computed is evaluated using appropriate tables. From this W^2, a modified W^2 can be calculated, and this is evaluated using a special table. With the standard table sample, sizes as low as 2 can be evaluated. Usable for individual data or those given as frequency distributions. | ST 70, ST 74, CO 71, MA 51, OW 62, ST 56 |

159

Table 4.2. (*Continued*)

Test Method	Computing Formula	Symbols Used	Remarks	Refs.
Watson	$U^2 = W^2 - n(S^*(x) - \frac{1}{2})^2$	Symbols as given, and $S^*(x) = S(x_i)/n$	This is a modification of the Cramer–von Mises method. The computed U^2 is modified, and this modified U^2 is then evaluated using a special table. Usable for individual data or those given as frequency distribution.	ST 70, ST 74
Anderson–Darling	$A^2 = -\{\sum (x_i - 1)[\ln S(x_i)] + (\ln[1 - S(x_{n+1-i})])]\}/n - n$	Symbols as given	The data are arranged in ascending order. The A^2 computed is modified and this modified value is evaluated using a special table. Usable for individual data or those given as frequency distributions.	ST 70, ST 74
Shapiro–Wilk	For even numbers of data: $n = 2k$ $$W = \frac{[\sum_1^k a_{n-i+1}(x_{n-i+1} - x_i)]^2}{(x_i - x_m)}$$ For odd numbers of data: $n = 2k + 1$ $$W = \frac{[a_n(x_n - x_1) + \cdots + a_{k+2}(x_{k+2} - x_k)]^2}{\sum (x - x_m)^2}$$	x_i = data $x_m = x_i/n$ a_n = special coefficients	The data are arranged in ascending order. The special coefficients needed for the calculation and the table needed to evaluate the W computed are accessible only via the references listed. Samples as low as 3 can be evaluated.	SH 65, SH 68, AG 71

Source: Reprinted with permission from Ames and Szonyi (AM 77). Copyright 1977 American Chemical Society.
[a]The various tests for normality have been reviewed by Shapiro et al. (AG 71, SH 65, SH 68, SH72), Schuster (SC 73), Dyer (DY 74), Klimko and Antle (KL 75), and Govindarajulu (GO 76).

160

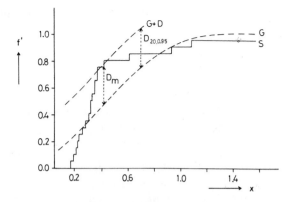

Figure 4.1. Cumulative frequency for Kolmogorov test.

3. Draw two lines parallel to the line of step 2, each at a distance D; D can be obtained from Table 4.3.
4. Draw the cumulative frequency line of the ordered data (S in Figure 4.1).
5. If S is found somewhere outside the lines of step 3, the distribution

Table 4.3. D Values for Different Levels of Significance α

n	$\alpha = .90$	$\alpha = .95$	$\alpha = .99$
1	0.950	0.975	0.995
2	0.776	0.842	0.929
3	0.642	0.708	0.828
4	0.564	0.624	0.733
5	0.510	0.565	0.669
6	0.470	0.521	0.618
7	0.438	0.486	0.577
8	0.411	0.457	0.543
9	0.388	0.432	0.514
10	0.368	0.410	0.490
12	0.338	0.375	0.450
14	0.314	0.349	0.418
16	0.295	0.328	0.392
18	0.278	0.309	0.371
20	0.264	0.294	0.356
> 20	$1.22/\sqrt{n}$	$1.36/\sqrt{n}$	$1.63/\sqrt{n}$

can be considered not normal with a probability given by the chosen value of Table 4.3).

Numerical Kolmogorov Test

1. Arrange the data in ascending order.
2. Compute $D_i = S(x_i) - i/n$, with i = sequential data points or frequencies, n = total number of data points, and $S(x_i)$ = corresponding normal cumulative distribution values.
3. Select the maximum value (D_m) of D_i.
4. If D_m is larger than the value given by Table 4.3, the distribution can be considered as not normal, with a probability given by the selected α-value of Table 4.3.

An example of both methods is given in Figure 4.1 and Table 4.4.

Table 4.4. Numerical Kolmogorov Test[a]

x_i(ordered)	i	i/n	$S(x_i)$	$|D_i|$
0.17	1	0.05	0.221	0.171
0.19	2	0.10	0.239	0.139
0.21	3	0.15	0.255	0.105
0.22	4	0.20	0.264	0.064
0.23	5	0.25	0.274	0.024
0.26	6	0.30	0.302	0.002
0.28	7	0.35	0.323	0.027
0.31	8	0.40	0.352	0.048
0.32	9	0.45	0.363	0.087
0.32	10	0.50	0.363	0.137
0.33	11	0.55	0.374	0.176
0.35	12	0.60	0.397	0.203
0.35	13	0.65	0.397	0.253
0.37	14	0.70	0.417	0.283
0.37	15	0.75	0.417	0.333[b]
0.42	16	0.80	0.476	0.324
0.61	17	0.85	0.677	0.173
0.93	18	0.90	0.915	0.015
1.08	19	0.95	0.963	0.013
1.57	20	1.00	0.999	0.001

[a] $m = 0.445$; $\sqrt{\beta_1} = 1.93$; $s = 0.354$; $D_m = 0.333$.
[b] Maximum value of D_i.

4.1.2. Transformation of Data

Occasionally one encounters a distribution that, by one of the tests mentioned above, departs too far from normality. It would not be safe then to apply the common statistical tests, and it would be a great advantage if, by a simple transformation of the variable, an approximately normal distribution could be obtained. For example, if the weights of a collection of spheres were given and found not to be normally distributed a natural transformation would be to take the cube root of the weight. Here one tests for the diameters to be normally distributed. Often a skewness may be removed or reduced by using the logarithm of the observation. The new distribution will not necessarily be normal but will probably pass the tests for normality. Some other typical transformations to normalize a skew distribution use the square, the square root, or the reciprocal of the variable.

Box and Cox (BO 64) described a universally applicable transformation for nonnormal data sets to create sets that are as near to normality as possible. For theoretical reasons they prefer to write:

$$y_i^{(\lambda)} = \frac{(x_i + c)^{\lambda} - 1}{\lambda} \qquad \text{for } \lambda \neq 0 \qquad (4.3)$$

$$y_i^{(\lambda)} = \log(x_i + c) \qquad \text{for } \lambda = 0 \qquad (4.4)$$

where x_i = the measured quantity
$y_i^{(\lambda)}$ = the transformed datum
λ = the transformation parameter
c = constant

For practical applications this can be reduced to

$$y_i^{(\lambda)} = (x_i)^{\lambda} \qquad \text{for } \lambda \neq 0 \qquad (4.5)$$

$$y_i^{(\lambda)} = \log x_i \qquad \text{for } \lambda = 0 \qquad (4.6)$$

The transformation is applied for a range of values of λ and each time a test for normality is performed. The best value of λ is chosen.

Box and Cox apply a somewhat different criterion in situations where an analysis of variance should be performed on the data. Note that the transformation rules mentioned above are implied in the often used "intuitive" transformations:

λ	Transformation
2	Square
1	Unaltered
0.5	Square root
0.333	Cube root
0	Log
-1	Reciprocal

In Figure 4.2 an example is given of data transformation and the various test results obtained. The original data are those of Table 4.4.

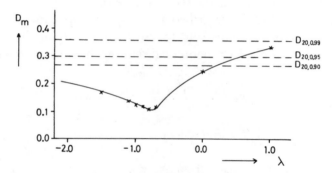

Figure 4.2. Estimation of optimal transformation factor.

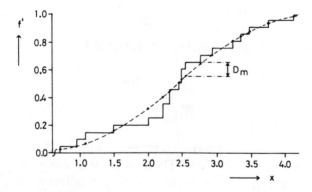

Figure 4.3. Kolmogorov test on transformed data of Figure 4.1.

As shown in Figure 4.2, the values D_m of the Kolmogorov test have a minimum at $\lambda = -0.8$. This means that a transformation according to $y_i^{(\lambda)} = (x_i)^{-0.8}$ gives a distribution that, according to the Kolmogorov test, is as near as possible to normal (see Figure 4.3 and Table 4.5). However, the values of $D_{n,\alpha}$ are such that, for example, for $D_{20,0.95}$ all distributions obtained by values of λ between $-2.5 < \lambda < 0.5$ should be considered as normal. Thus an "intuitive" transformation by taking the square root ($\lambda = 0.5$), the logarithm ($\lambda = 0$), or the reciprocal ($\lambda = -1$) would suffice as well.

Table 4.5. Transformation of Data to a Normally Distributed Data Set

$\lambda = 1$	$\lambda = -0.8$
x_i	y_λ^i
0.17	4.127
0.19	3.776
0.21	3.485
0.22	3.358
0.23	3.241
0.26	2.938
0.28	2.769
0.31	2.552
0.32	2.488
0.32	2.488
0.33	2.428
0.35	2.316
0.35	2.316
0.37	2.215
0.37	2.215
0.42	2.002
0.61	1.485
0.93	1.060
1.08	0.940
1.57	0.697
$m = 0.445$	$m = 2.445$
$s = 0.354$	$s = 0.918$
$\sqrt{\beta_1} = 1.93$	$\sqrt{\beta_1} = -0.177$
$D_m = 0.333$	$D_m = 0.102$

4.2. INFORMATION THEORY

According to a definition of Gottschalk (GO 72), "analytical chemistry provides us with optimal strategies for collecting and appreciating relevant information on situations and processes in materials." The input into the analytical procedure contains some actual information and some latent information (uncertainty). During the analytical process the actual information increases at the expense of the latent information. For example, a sample is analyzed for iron content. The actual information here is that iron is expected to be present and the amount is to be determined. The latent information is the unknown quantity of iron. By an analytical procedure the iron content is measured and found to be $10.82 \pm 0.01\%$ weight. Thus the uncertainty is decreased from a range of $0-100\%$ to that of $10.81-10.83\%$ at the expense of the latent information.

Because the output is information that is disclosed by measurements, the definition of this quantity needs a closer examination. One definition stems from information theory (SH 49): "Information equals diminishing uncertainty." In our example the iron content before analysis equals $0-100\%$ and afterward $10.81-10.83\%$. Using a more general definition we can state that information I equals uncertainty $H_{before} - H_{after}$ analysis. Another name for uncertainty H is entropy. The magnitude of H is given by the number of possibilities (n) before and after the experiment.

The most simple experiment is a decision between two alternatives, yes or no. The yield of information of such an experiment is selected as a unit of information, the bit (binary unit). This selection implies

$$I = k \log_z 2 = 1 \text{ (bit)} \tag{4.7}$$

and $k = 1/\log_z 2$. Thus

$$I = \frac{\log_z n}{\log_z 2} = \log_2 n = \text{ld } n \tag{4.8}$$

Here ld represents the logarithm with base 2. Other units found in literature are "nit," based on logarithm with base e, and "dit," based on information obtained by selecting 1 out of 10 independent possibilities.

If a selection must be made of 1 out of n possibilities with unequal possibilities, the equation of Shannon (eq. 4.9) (SH 49) should be applied:

$$I = H = - \sum_{i=1}^{n} P_i \operatorname{ld} P_i \qquad (4.9)$$

A possibility to be mentioned is a probability distribution represented by a Gaussian function. Such a distribution may be approximated by means of a histogram with a given width of the classes Δx. The smaller is Δx, the more classes are needed and the better the approximation. The probability of occurrence of a given class equals

$$\rho_i = \int_{x_i}^{x+\Delta x} \rho(x_i)dx = \rho(x_i)\Delta x \qquad (4.10)$$

The entropy of a real continuous distribution may be approximated by the entropy of the histogram

$$H = - \sum_{i=1}^{n} P_i \operatorname{ld} P_i = - \sum_{i=1}^{n} [P(x_i)\Delta x] \operatorname{ld}[P(x_i)\Delta x] \qquad (4.11)$$

4.2.1. Application of Information Theory to Quantitative Analysis

A good introduction is given by Eckschlager (EC 85). The constant P_c of a certain compound in a mixture before and after analysis is depicted in Figure 4.4. Before the analysis each concentration c_0 and c_e has an equal probability of occurrence; after analysis the concentration is known with a deviation of δ_c.

Application of the Shannon equation to the situation before analysis yields

$$
\begin{aligned}
H_{\text{before}} &= - \int_{-\infty}^{+\infty} P_c(\operatorname{ld} P_c)dc - \operatorname{ld} \Delta x \\
&= - \int_{-\infty}^{c_0} P_c(\operatorname{ld} P_c)dc - \int_{c_0}^{c_e} P_c(\operatorname{ld} P_c)dc \\
&\quad - \int_{c_e}^{+\infty} P_c(\operatorname{ld} P_c)dc - \operatorname{ld} \Delta x \\
&= 0 - P_b(c_e - c_0)\operatorname{ld} P_b - 0 - \operatorname{ld} \Delta x
\end{aligned}
\qquad (4.12)
$$

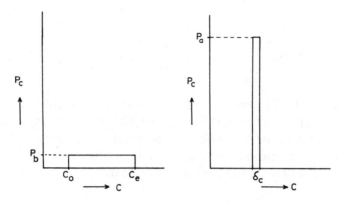

Figure 4.4. Probability of occurrence of a concentration P as a function of the numerical value of that concentration C before (P_b) and after (P_a) analysis.

and after analysis

$$H_{\text{after}} = 0 - P_e(\text{ld } P_e)\delta_c - \text{ld } \Delta x \qquad (4.13)$$

Before analysis

$$\int_{c_0}^{c_b} P_b dc = 1 \qquad (4.14)$$

or

$$P_b = \frac{1}{c_e - c_0} \qquad (4.15)$$

which on substitution in eq. 4.12 yields

$$\begin{aligned} H_{\text{before}} &= \frac{1}{c_e - c_0}[ld(c_e - c_0)](c_e - c_0) - \text{ld } \Delta x \\ &= \text{ld}(c_e - c_0) - \text{ld } \Delta x \end{aligned} \qquad (4.16)$$

After analysis $P_e\delta_c = 1$ and thus

$$H_{\text{after}} = \text{ld } \delta_c - \text{ld } \Delta x \qquad (4.17)$$

The information gain I becomes

$$I = H_{\text{before}} - H_{\text{after}}$$
$$= \text{ld}\left(\frac{c_e - c_0}{\delta_c}\right) \tag{4.18}$$

Application of this procedure in a situation giving a Gaussian distribution of the concentration after analysis implies

$$P_c = \frac{1}{\sigma\sqrt{2\pi}}\exp\left[-\frac{(c-\mu)^2}{2\sigma^2}\right] \tag{4.19}$$

in which σ is the standard deviation and μ the position of the maximum probability. This distribution results in the following H_a value:

$$H_a = \text{ld}(\sigma\sqrt{2\pi e}) - \text{ld}\,\Delta x \tag{4.20}$$

and thus

$$I = \text{ld}\left(\frac{c_e - c_0}{\sigma\sqrt{2\pi e}}\right) \tag{4.21}$$

In situations where σ is unknown, its value may be estimated from n_s measurements and the mean result of the measurements $\bar{c}$:

$$s = \left[\frac{\sum\limits_{i=1}^{n_s}(c_i - \bar{c})^2}{n_s - 1}\right]^{1/2} \tag{4.22}$$

A somewhat more general expression is given by

$$I = \text{ld}\frac{(c_e - c_0)\sqrt{n_s}}{2t_{p,v}\,s} \tag{4.23}$$

Here $t_{p,v}$ represents the probability distribution of the analytical results p and a parameter v, representing the number of degrees of freedom.

Thus the gain of information depends on s and n. It is readily seen (eq. 4.23) that an increase in the number of measurements is not proportional to the increase of information: the increase of information is less than the information expected from the number of measurements. The diminishing effect of the estimated number of measurement on information gain is called redundancy R:

$$
\begin{aligned}
R &= I_{max} - I_n \\
&= nI - I_n
\end{aligned}
\tag{4.24}
$$

This equation implies that the maximum amount of information to be gained from n measurements equals n times the information of one single measurement. The interaction between successive measurements thus is supposed to be absent. In the case that correlation exists between measurements

$$
I = I + \tfrac{1}{2}\mathrm{ld}\,|\mathrm{corr}|
\tag{4.25}
$$

where corr = correlation matrix.

The redundancy provides a method for estimation of the mutual interaction of successive measurements. A high value of the redundancy implies that the effects of interruptions otherwise leading to erroneous results are partially canceled by the correct results in a series of measurements. In quantitative analysis a calibration graph frequently is used. The question then arises: how is the information theory applied here? To begin with, the standardized function $u = ac + b$ is constructed from m samples with concentrations c_i and measurements u_i. The analysis of a sample yields a measurement u_a and thus a concentration c_a. A set of n parallel measurements result in $\bar{u}_a$ and $\bar{c}_a$. With a linear dependence of c on u the error in c_a, Δc_a, equals

$$
\Delta c_a = \frac{t_{p,v}s_u}{a}\left\{\frac{1}{n} + \frac{1}{m} + \frac{(\bar{u}_a - \bar{u})^2}{a^2\left[\sum\limits_{i=1}^{m} c_i^2 - \left(\sum\limits_{i=1}^{m} c_i\right)^2\Big/m\right]}\right\}^{1/2}
\tag{4.26}
$$

The actual concentration equals $c_a \pm \Delta c_a$. Substitution into eq. 4.23 with $s_u t_{p,v}/\sqrt{n} = \Delta c_a$ results in

$$
I = \mathrm{ld}\left[\frac{(c_i - c_0)a}{2s_u t_{p,v}}\right] - \frac{1}{2}\mathrm{ld}\left\{\frac{1}{n} + \frac{1}{m} + \frac{(\bar{u}_a - \bar{u})^2}{a^2\left[\sum\limits_{i}^{m} c_i^2 - \left(\sum\limits_{i}^{m} c_i\right)^2\Big/m\right]}\right\}
\tag{4.27}
$$

This implies that the information increases upon

- A higher value of the sensitivity a of the analysis
- An increase of n, the number of duplicate measurements
- A decrease of s_u, which means an increase in reliability
- A higher number of testing samples m
- A higher range of combinations of the testing samples
- Measuring close to the center of gravity of the test set $(\bar{u}_a - \bar{u})$ becomes smaller.

A small difference between C_e and c_0 results in a decrease in information. This disadvantage may partly be compensated by centering the testing samples around the expected value of the analysis. The most important observation to be made from eq. 4.26 is the overriding effect of sensitivity a and reliability s_u in comparison with the number of test

Table 4.6. Derivation of the Optimal Number of Test Samples[a]

$c_3 - c_o = 100$

$\bar{u}_a - u = 0$

$s_u = 0.01$

$a = 0.5$

I (bit)	α in $t_{p,v} = 0.039$							
	$n=1$	$n=2$	$n=3$	$n=4$	$n=5$	$n=10$	$n=100$	$n=\infty$
$m=3$	7.04	7.40	7.54	7.63	7.70	7.85	8.02	8.04
$m=4$	8.83	9.20	9.38	9.49	9.56	9.75	9.96	9.99
$m=5$	9.33	9.72	9.91	10.04	10.12	10.32	10.59	0.62
$m=6$	9.60	9.99	10.21	10.34	10.43	10.66	10.96	11.00
$m=7$	9.73	10.14	10.35	10.50	10.60	10.84	11.18	11.23
$m=10$	9.92	10.36	10.60	10.75	10.80	11.15	11.59	11.65
$m=100$	10.19	10.68	10.99	11.17	11.32	11.79	13.01	13.65
$M \to \infty$[b]	10.24	10.74	11.03	11.24	11.40	11.90	13.56	—

[a] When a straight line calibration curve is employed the number of degrees of freedom v equals $(m-2)$.

[b] Here $m \to \infty$ holds when the calibration coefficient is known exactly, for example, because of a theoretical constant.

samples m or the number of duplicates n. Based on eq. 4.26 an optimal value of m, the number of test samples, can be derived for a given value of a, $c_e - c_0$, $\bar{u}_a - u$, n, s_u, and $t_{p,v}$ and the required amount of information in, for example, $I \geqslant 9.5$ (Table 4.6).

4.2.2. Application of Information Theory to Qualitative Analysis

The measurement of the identity of a given sample is a qualitative analysis. Cleij and Dijkstra (CL 79) treated the theory on the application of information theory. Here only some practical examples are given. In order to fix the identity of a compound, information is required, for example, the measurement of the melting point. The amount of information thus obtained depends on the probability of occurrence of that particular melting point.

In practice melting points cannot be measured exactly; therefore the range over which they are distributed is subdivided into a number of classes. The number of compounds with a melting point located in a given class gives the probability of occurrences of a particular melting point. To illustrate this a histogram has been constructed

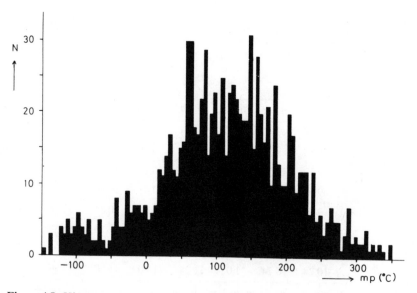

Figure 4.5. Histogram for melting point distribution of 1001 arbitrarily chosen compounds: Class width, 5 °C; range, −150 °C to +350 °C.

starting with 1349 arbitrarily chosen compounds. From this set the melting points of 348 compounds are unknown. The melting points of the other compounds are distributed over 100 classes with a class width of 5 °C ranging from − 150 °C up to + 350 °C (see Figure 4.5).

If we assume an accuracy of the melting point measurement corresponding with the width of the classes in the histogram, the information yield of a melting point measurement equals

$$I = -\sum p_i \, \text{ld} \, p_i$$
$$= 6.17 \, (\text{bit})$$

(4.28)

A more accurate measurement of the melting point results in a relatively smaller reliability interval in comparison with the width of a class and correspondingly provides more information. In order to account for this effect we assume an accuracy for the melting points equal to k times that of the histogram intervals. The melting point measurement may be considered as a combination of two experiments:

1. A class selection procedure with an information yield of $I_1 = \sum_{i=1}^{n} p_i \, \text{ld} \, p_i$
2. Within a class selection of 1 possibility out of a set of k, with an information yield of $I_2 = \text{ld} \, k$, or in terms of class broadness Δx and reliability interval Δy, $I_2 = \text{ld} \, \Delta x / \Delta y$

The total amount of information thus becomes

$$I = -\sum_{i=1}^{n} P_i \, \text{ld} \, P_i + \text{ld} \, \frac{\Delta x}{\Delta y}$$

(4.29)

This calculation implies equal probability for all k possibilities within a given class. In many situations this is not true. In order to deal with situations where the probability of the situations are not equal, a probability distribution function within a class may be defined.

The equations involved in such situations are

$$I'_1 = -\sum_{i=1}^{n} P_i \, \text{ld} \, P_i$$

(4.30)

for information obtained of intraclass distribution and

$$I'_2 = \left(-\sum_{j=1}^{k} q_j \operatorname{ld} q_j \right) = \left(-\sum_{j=1}^{\Delta x/\Delta y} q_j \operatorname{ld} q_j \right)_i \qquad (4.31)$$

For information of intraclass distribution in class i, q_j equals the probability of occurrence of the measured melting point in graph j of class i.

The average intraclass information $\bar{I}'_2$ for the entire histogram equals

$$\bar{I}'_2 = \frac{1}{n} \sum_{i=1}^{n} \left(-\sum_{j=1}^{\Delta x/\Delta y} q_j \operatorname{ld} q_j \right)_i \qquad (4.32)$$

which makes the total information $I_{tot} = I'_1 + \bar{I}'_2$:

$$I_{tot} = -\sum_{i=1}^{n} P_i \operatorname{ld} P_i + \frac{1}{n} \sum_{i=1}^{n} \left(-\sum_{j=1}^{\Delta x/\Delta y} q_j \operatorname{ld} q_j \right)_i \qquad (4.33)$$

In the special situation where q_i has a constant value $\Delta y/\Delta x$, eq. 4.33 simplifies to

$$\begin{aligned} I_{tot} &= -\sum_{i=1}^{n} P_i \operatorname{ld} P_i + \operatorname{ld} \frac{\Delta x}{\Delta y} \\ &= -\sum_{j=1}^{m} P'_j \operatorname{ld} P'_j \end{aligned} \qquad (4.34)$$

where $m = n\Delta/\Delta y$ and $P'_j = \Delta x$.

Equation 4.34 refers to the distribution of melting point over a new histogram with class width m. Based on the histogram and an accuracy of $1\,^\circ C$ in the measurement of the melting points, an amount of information may be obtained

$$I_{tot} = 6.17 + \operatorname{ld} \frac{5}{1} = 8.5 \,(\text{bit}) \qquad (4.35)$$

Now one may ask how to use this information and whether this is enough? Based on an estimated number of 10^6 compounds, the identification of one particular compound requires an amount of information $I = \mathrm{ld}\, n_0/n_1$, which here means $\mathrm{ld}\, 10^6/1 = 20$ bits. The amount of information obtained from the melting point, namely, 8.5 bits, reduces the number of possibilities by $2^{8.5} \approx 360$.

Another useful identification method is thin-layer chromatography (TLC) (MA 73). Information theory can be applied to this method in order to find the best separation and thus the maximal possible identification. This may be illustrated by an example. In order to identify one particular compound out of a set of 8, an amount of information of $\mathrm{ld}\, 8 = 3$ bits is required.

A theoretical solution for this identification problem is provided by two separating schemes, one giving 2 bits of information and the other 1 bit. In practice this solution is not always possible because part of the information provided by the first solvent may be identical with that given by the second solvent, which results in a total gain of information of less than 3 bits. Therefore most information is obtained from a combination of two separating solvents with the highest possible independent information yield.

This can be illustrated by the example given in Table 4.7. It is concluded from the R_F values that the individual liquids do not provide enough information for the identification of the compounds $a-h$. Thus the amount of information is enhanced by application of

Table 4.7. R_F **Values of Eight Examples for Three Different Separating Liquids (MA 73)**

Compound	Liquid 1	Liquid 2	Liquid 3
a	0.20	0.20	0.20
b	0.20	0.40	0.20
c	0.40	0.20	0.20
d	0.40	0.40	0.20
e	0.60	0.20	0.40
f	0.60	0.40	0.40
g	0.80	0.20	0.40
h	0.80	0.40	0.40
Amount of information	2 bits	1 bit	1 bit
$-P_i\,\mathrm{ld}\,P_i$	$(\log_2 4)$	$(\log_2 2)$	$(\log_2 2)$
	$-4 \cdot \frac{1}{2}\,\mathrm{ld}\,\frac{1}{2}$	$-2 \cdot \frac{1}{4}\,\mathrm{ld}\,\frac{1}{4}$	

Source: Reproduced from D. L. Massart, *J. Chromatogr.* **79**, 157 (1973) with permission of Elsevier Scientific Publishing Company, Amsterdam.

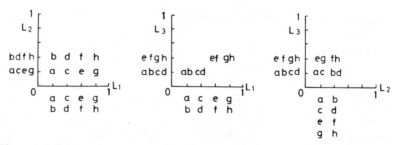

Figure 4.6. Schematic representation of two successive separations with different separating liquids L_1, L_2, and L_3 for three different selections of liquids.

two successive separations with different separating liquids. There are various combinations of liquids possible. From Figure 4.6 one concludes that the combination of L_1 with L_3 provides 2 bits of information only, because the information provided by L_3 has been given already by L_1. The same conclusion holds for the combination L_2 and L_3. The only combination that provides all information required for identification is L_1 with L_2.

Selecting appropriate separating liquids for separation of larger members of compounds follows the same line in principle. Mostly the interdependence of the information provided follows from the ratio of R_F values of compounds for the different separating liquids. When the R_F values obtained from two different separating liquids are highly correlated, the information gain in combined separation is small and no extra separation is obtained. A low correlation provides better separation and more information on application of two separations.

4.3. FILTERING

In order to apply measured quantities, in most situations the following treatment is required:

- Removal of nonessential—irrelevant—data (noise)
- Reconstruction of essential data
- Interpretation of data

These treatments are discussed in the following subsections.

4.3.1. Removal of Irrelevant Data (Noise)

A practical definition of noise is that it is the undesired part of a signal. In practice it is impossible to separate noise completely from the rest of the signal because each noise filter generates noise itself. A practical yardstick for the applicability of a signal is provided by the signal-to-noise (S/N) ratio. In order to measure this ratio one starts by measuring the available power of the signal wave and of the noise wave. In direct current signals the power is proportional to the voltage E; in alternating current signals it is proportional to the root mean square value A_{rms} of the waveshape:

$$A_{rms} = \left[\sum_i^K \frac{(\bar{A} - A_i)^2}{K} \right]^{1/2} \tag{4.36}$$

with $\bar{A}$ being the average value of the waveshape, and A_i the deviation of the ith measurement from the average.

With direct current signals sampled at discrete time intervals the main part of the noise is governed by a variation with time of the signal. Here the noise is given by the standard deviation of the signal and thus the signal-to-noise ratio becomes

$$\frac{S}{N} = \frac{\text{average value}}{\text{standard deviation}} = \frac{1}{\text{relative standard deviation}} \tag{4.37}$$

Equation 4.37 gives a simple method to calculate S/N from a recorder signal; for example, 99% of all observations is found between $\mu - 2.5\sigma$ and $\mu + 2.5\sigma$ (μ equals the main value, and σ denotes the estimated standard deviation). Here S/N equals 0.2, the peak-to-peak variance of the signal. The observation period should be long enough to enable a reliable estimate of S. It is assumed that the signal is distributed around the average value μ according to a Gaussian distribution.

In situations where alternating current signal are measured, other techniques should be applied for calculating S/N, for example, autocorrelation (see Section 4.5.2.1).

One may distinguish various types of noise that can be recognized by their frequency spectrum (Figure 4.7);

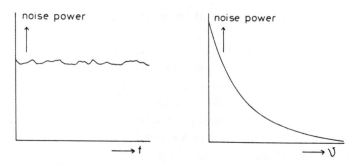

Figure 4.7. Frequency spectrum of "white" noise and of "$1/f$" noise.

- *White noise*—has a frequency spectrum with a flat shape. It is caused by random processes such as Brownian motion, external influence, and thermal motion of electrons (Johnson noise).
- *$1/f$ noise*—where the power is proportional to $1/f$. Such a situation is met with, for example, a drifting signal.
- *Red noise*—contains more low-frequency noise than the average value.
- *Blue noise*—contains more high-frequency noise than the average value.
- *Interference noise*—nearly always results from external sources with relatively low chance of occurrence, i.e., switching on or off of power, undesired components in a spectrum, and so on. A special form of interference noise is *impulse noise*. It is caused by sources such as the starting up of a powerful engine. All frequencies are met in this type of noise, making it difficult to eliminate. Another type of interference noise is *deformation noise*. Here some specific frequencies are added to or removed from the signal. Some possible causes of deformation noise are nonlinearity of the instrument, transmission losses, and reflections.

4.3.2. Data Filtering

In practice all types of noise occur together, giving a mixture. The noise energy generally depends on $1/f^n$, where $0 < n < 1$.

Removal of noise from a signal is performed by filtering. In order to construct a good filter, one starts with fixing the desired frequency or frequencies; based on these, a filter is designed that allows passage of frequencies in a very narrow range around the selected one.

This is done by means of smoothing: the signal is integrated over an extended period in order to cancel the higher frequencies.

4.3.2.1. Digital Filtering

Digital smoothing requires determination of the running average of a series of measurements x_t as a function of time:

$$S_t = \frac{1}{N}(x_t + x_{t-1} + \cdots + x_{t-N+1}) \tag{4.38}$$

or

$$S_t = S_{t-1} + \frac{1}{N}(x_t - x_{t-N}) \tag{4.39}$$

where S_t equals the running average at time t, and N equals the time span of the observation period expressed in time units.

Equation 4.39 describes how to calculate a new S_t value from an older S_{t-1} by adding a new measured quantity x_t and eliminating the oldest x_{t-N}. This procedure asks for greater memory capacity because a complete series of N data should be stored.

Exponential smoothing requires considerably less storage capacity because nonavailable data are substituted by the mean value: in Eq. 4.39 $(x_t - x_{t-N})/N$ is replaced by $(x_t - S_{t-1})/N'$. Here S_t equals the running exponential mean at time t, and N' the mean length of the averaging period. Substitution of $\alpha = M'^{-1}$ in eq. 4.39 yields

$$S_t = \alpha x_t + (1 - \alpha) S_{t-1}$$
$$= S_{t-1} + \alpha(x_t - S_{t-1}) \tag{4.40}$$

It can be seen that the new S_t value is obtained from the old by correcting with the scaled difference between the old value and the most recently measured value. The scaling factor α is inversely proportional to the length of the running interval. The advantage of this method over the previous one is that only the mean value S_{t-1} has to be stored in memory; moreover, all measured values are accounted for, in contrast to the running average method. The importance of a measurement decreases with its age; this is of consequence for the mean age $\bar{k}$ of a series of measurements (BR 62). The age of the most recent measurement equals 0, the next to most recent one equals 1, and so on. Thus the average age $\bar{k}$ equals

$$\bar{k} = \frac{1-\alpha}{\alpha} \tag{4.41}$$

The running average method has an average age of the observations for a period N:

$$\bar{k} = \frac{0+1+2+3+\cdots+(N-1)}{N} = \frac{N-1}{2} \tag{4.42}$$

Equalizing these two ages yields

$$\frac{1-\alpha}{\alpha} = \frac{N-1}{2} \tag{4.43}$$

or

$$\alpha = \frac{2}{N+1} \tag{4.44}$$

The weight factor depicted as a function of time is given in Figure 4.8.

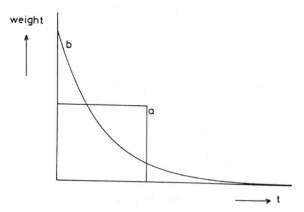

Figure 4.8. Weight factor as a function of time for running average (a) and for exponential smoothing (b).

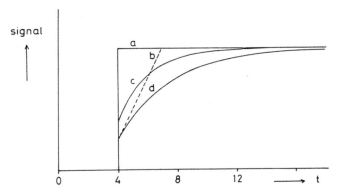

Figure 4.9. Modification of input signal (a) as a function of time with the running average method with $N = 4$ (b), with an exponential with $\alpha = \frac{1}{4}$ (d), and with an exponential $\alpha = 2/N + 1 = 0.4$ (c).

The shape of a signal is altered by smoothing and depends on the smoothing technique applied. It is shown in Figure 4.9 that the shape of the signal depends on the smoothing procedure irrespective of whether an equal time span for the observations is applied.

Frequency filtering for one specific frequency v_0 allows only v_0 to pass the filter. All time-dependent changes of the signal are lost. Therefore the filter should allow a well-defined bandwidth. Non-periodic but repeatable signals can be noise filtered by averaging. Signal series should be sampled with a frequency at least twice the highest occurring frequency in order to describe the signal series completely (see Section 4.3.2.3). In order to average signal series in a well-defined way, synchronization should be applied. The synchronization pulse should be deferred from the signal itself or should trigger the signal registration. This method has been elegantly applied by Smit (SM 80), who describes a method for pseudorandom injections of a sample into a chromatographic column and cross-correlating the detector signal with the noise generator that governs the injections. With this method it is possible to accumulate a signal from a detector output that under normal conditions is hidden in the background noise.

A method for producing synchronization pulses from the signal itself consists of monitoring the sign of the derivative and selecting the moments when the sign equals zero after passing from negative toward positive values.

Digital signal averaging is a commonly used technique that requires as many channels as signals to be averaged. Interference noise is not averaged out by signal averaging.

In situations where synchronization pulses are not available, correlation techniques can be useful. Here the signal value at a moment t is multiplied with a signal value an interval τ apart; thus

$$\varphi_{zz}(\tau) = \sum_{t=1}^{N-\tau} \frac{(z_t - \bar{z}_t)(z_{t+\tau} - \bar{x}_{t+\tau})}{(n-\tau)\sigma_z^2} \qquad (4.45)$$

(for a stationary signal $\bar{z}_t = \bar{z}_{t+\tau}$). The correlated signals become clearly distinguishable because noise is eliminated. The correlation time may be used to describe the signal. Autocorrelating a sine function produces a cosine function. It is known that $\tau = 0$ produces the variance σ^2 of the signal; it consists of signal and noise.

4.3.2.2.   Kalman Filtering

A measured quantity y_t, obtained at time t, is just a momentary description of a situation that takes no account of the history of a time series of measurements. However, it is possible to make a better estimate of reconstructed values out of a time series at time t and to predict the value of new measurements taking into account the history of that time series. Here the prediction of a measured value is based on the most recent measurement, and the expected value is corrected with the values measured in the more distant past. The weight of the predicting capacity of the measured values from the past decreases with increasing distance from the present time. The application of this technique is called exponential smoothing. Kalman (KA 61) showed that there exists an optimal estimation method that is even better than the exponential smoothing method.

Excellent overviews are those of Brown (BR 86) and Rutan (RU 89). The predicted value for time t, called $\bar{x}_t$, is combined with a measured value y_t in order to construct a better estimate called $\hat{x}_t$ (each of the data x_t and y_t contains errors, which makes an exact prediction impossible):

$$\hat{x}_t = (1 - k_t)\bar{x}_t + k_t y_t \qquad (4.46)$$

Analytical chemistry can be described by a set of very simple equations:

$$X_k = F_k X_{k-1} + W_{k-1} \qquad (4.47)$$

$$Z_k = H_k X_k + V_k \qquad (4.48)$$

Here X_k, often called the state of the system, can for instance be a concentration. Now eq. 4.47 describes the concentration at time (or sequence) k to be a function of the past; F_k is a transition function. In the case of a stable sample where the concentration will not change in time, F_k will be 1. In the case of a process, F_k can for instance be $F_k = \exp(-\alpha_k)$.

As the value of X_k will be between certain values we do not know, this is denoted by W_{k-1}, the system error. Equation 4.48 gives the relation between state (e.g., the concentration) and a measured value X_k. Here H_k can be the calibration factor, and V_k denotes the measurement error.

Now X_k can be predicted:

$$\bar{X}_k = F_k \hat{X}_{k-1} \qquad (4.49)$$

where $\bar{X}_k$ is the predicted value of X_k
         $\hat{X}_{k-1}$ is the estimated value of X_{k-1}, obtained in the past

Next we can make an estimate $\hat{X}_k$ of X_k (it can be proven that this is the optimal estimate) by making a linear combination of the predicted value $\bar{X}_k$ and the measured value Z_k:

$$\hat{X}_k = (1 - K_k)\bar{X}_k + K_k Z_k \qquad (4.50)$$

where K_k is a weight factor. In fact eq. 4.50 tells us not to rely only on the last measurement Z_k but also to use the experience from the past. The weight factor K_k, determining the relative weights of the present and history, can be estimated from

$$K_k = \bar{P}_k (\bar{P}_k + R_k)^{-1} \qquad (4.51)$$

where $\bar{P}_k$ is the variance in the system, and R_k is the variance of the measurement ($R_k = V^2$). So a high value of R_k (a high measurement noise) sets K_k to be low and the values from the past get a higher weight through eq. 4.50.

A larger uncertainty in the system (P_k high) gives more weight to the last measurement. Then $\bar{P}_k$ can be predicted from

$$\bar{P}_k = F_k^2 \hat{P}_{k-1} + W_{k-1}^2 \tag{4.52}$$

and P_k can be estimated from

$$\hat{P}_k = (1 - K_k)\bar{P}_k \tag{4.53}$$

As this set of equations is recursive a starting value for X_k and P_k is required. These can be guessed from past experience or, when such experience is lacking, X_k can be set to 0, P_k to a high value.

As the advantage of the Kalman filter is the exploitation of the knowledge obtained in the past and updating this knowledge continuously, the method can be used to enhance the quality of analytical measurements.

The set of equations in matrix notation from Table 4.8 can be used to apply the Kalman filter to more general problems.

Thijssen (TH 85) applied a Kalman filter to calibration control. For a linear calibration graph with parameters drifting at random, the accompanying model has been derived (eq. 4.54) (TH 84).

$$Z_k = (c, 1) \begin{vmatrix} a(k) \\ b(k) \end{vmatrix}$$

$$X_k = F(k-1)X_{k-1} + W_{k-1}$$

$$\begin{vmatrix} a(k) \\ b(k) \\ \alpha(k) \\ \beta(k) \end{vmatrix} = \begin{vmatrix} 1 & 0 & 1 & 0 \\ 0 & 1 & 0 & 1 \\ 0 & 0 & 1 & 0 \\ 0 & 0 & 0 & 1 \end{vmatrix} \begin{vmatrix} a(k-1) \\ b(k-1) \\ \alpha(k-1) \\ \beta(k-1) \end{vmatrix} + \begin{vmatrix} W_1(k-1) \\ W_2(k-1) \\ W_3(k-1) \\ W_4(k-1) \end{vmatrix} \tag{4.54}$$

$$Z_k = |c, 1, 0, 0| \begin{vmatrix} a(k) \\ b(k) \\ \alpha(k) \\ \beta(k) \end{vmatrix} + V(k)$$

Table 4.8. Equations for Kalman Filtering

Model:

$$X_k = F(k, k-1)X_{k-1} + W_{k-1}$$

$$Z_k = H_k^t X_k + V_k$$

Prediction:

$$\hat{X}_{k/k-1} = F(k, k-1)\hat{X}_{k-1/k-1}$$

$$P_{k/k-1} = F(k, k-1)P_{k-1/k-1}F^t(k, k-1) + Q_{k-1}$$

Filtering:

$$\hat{X}_{k/k} = \hat{X}_{k/k-1} + K_k[Z_k - H_k^t \hat{X}_{k/k-1}]$$

$$P_{k/k} = P_{k/k-1} - K_k H_k^t P_{k/k-1}$$

$$K_k = P_{k/k-1}H_k[H_k^t P_{k/k-1}H_k + R_k]^{-1}$$

where for $k = 0$, $\hat{X}_{k/k} = 0$ and $P_{k/k} = I_n sO^2$

The *state* refers to the parameters of a linear calibration graph $z = aC + b$, where z is the signal and C the concentration. In order to model drift with time, the slope a and intercept b are extended by α and β, respectively. The state is estimated by measuring standards with known concentrations or reference values, which are processed by the Kalman filter. For the evaluation of unknown samples, the associated signal and the predicted state are used in the formulation $C = (z - b)/a$. After each measurement, a quality control algorithm decides whether to calibrate again or to process the next unknown. For evaluation, a preselected criterion N_{crit} is compared with the maximal relative imprecision N_{max} of the expected results. The decision to recalibrate is followed by selecting which of the available concentration standards gives the optimal performance of the Kalman filter.

The calibration system with algorithms for prediction, filtering, evaluation, control, and optimization operates on-line and the results can be used directly for analytical purposes. After an entire batch of samples has been processed and if a time delay is not detrimental, the preset quality of the results can be improved by smoothing. Application of the Kalman procedure to a four-dimensional filtering problem is given by Poulisse (PO 79). He uses a multicomponent analysis computation on spectroscopic data and compares this method of calculation with a nonrecursive least-squares estimation method (see Section 4.4). His calculations result in a determination of the number of components in the mixture, an estimation of a constant systematic error, and a computation of the error covariance matrix. In this paper theory and practice both point to a new design of measurement experiments and novel applications of on-line computation and computer control. It is shown that the Kalman filter algorithm, because of its recursive nature and computational efficiency, is a very convenient method for multicomponent analysis computations and offers promising possibilities in acquiring a predetermined quality of the results in analytical chemistry.

Another application of Kalman filtering is presented by Didden and Poulisse (DI 80). Here a Kalman filter is designed to determine the number of components in multicomponent mixtures. Optical spectra [ultraviolet–visible range (UV–vis)] of different mixtures are simulated and processed sequentially by the Kalman filter. It is shown in this study that the number of components present in the sample as well as their concentrations can be determined simultaneously. The advantages of the method in view of the computational effort required are demonstrated.

4.3.2.3. Fourier Convolution

Here a time series is written as a summation of a number of sine and cosine terms. This procedure is allowed for each type of function $f(x)$:

$$f(x) = a_1 \sin x + a_2 \sin 2x + a_3 \sin 3x + \cdots + b_0 + b_1 \cos x$$
$$+ \cdots + b_n \cos nx \tag{4.55}$$

In order to calculate the coefficients a_m, $f(x)$ is multiplied by $\sin mx \, dx$ ($m = 1, 2, \ldots$) and integrated between $-\pi$ and π. This process is visualized as shown in Figure 4.10. It is seen that even functions "$f(x) = +f(-x)$" make all a coefficients zero and odd functions "$f(x) = -f(-x)$" make all b coefficients zero.

In situations where the periodicity of $\sin mx$ corresponds with a periodicity of $f(x)$, the product function does not change sign because a change in sign of $\sin mx$ occurs at the same x value where a change in

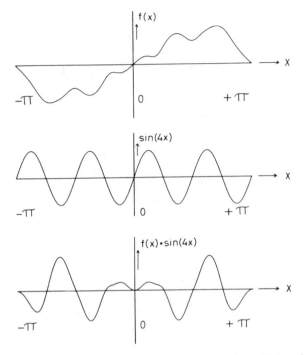

Figure 4.10. Schematic representation of the convolution of an odd function $f(x)$ with the sine function $\sin(4x)$.

sign of $f(x)$ occurs. All periods of the product function contribute with the same algebraic sign to the integral and the result is a nonzero value of a_m. In situations where the periodicity of $\sin mx$ does not correspond with a periodicity of $f(x)$ the product function tends to be plus as often as minus and the contributions of the periods to the integral tend to cancel each other out. The result is a practical zero value for a_m.

The same procedure is repeated with $\cos mx$ ($m = 0, 1, 2, \ldots$) in order to find coefficients b_m.

Generally a series contains a limited number of terms. Where the highest frequency of the series equals mx, the base frequency consists of m terms, and thus the series consists of m cosine functions, m sine functions, and one constant, which means $2m + 1$ terms. The amplitudes are derived from $2m + 1$ measurements, which means that the number of measurements (minimal and in practice also maximal) equals about $2m$.

The sampling has to provide $2m$ samples, which at a constant sampling frequency v means an intersample time span Δt:

$$\Delta t = \frac{v}{2m} = \frac{v}{2v} \tag{4.56}$$

where v equals the length of the series expressed in time units. The shape of a wave may be characterized by a number of sine and cosine functions. An even function requires a description with exclusively cosine functions, for example, $f(x) = -\frac{1}{4}\pi$ with $0 < x < \pi$; $f(x) = 0$ with $x = 0$ and $f(x) = \frac{1}{4}\pi$ with $-\pi < x < 0$. Here $f(x)$ is an odd function and the series is formed by a summation of sine terms exclusively (Figure 4.11):

$$f(x) = a_1 \sin x + a_2 \sin 2x + a_3 \sin 3x + \cdots \tag{4.57}$$

$$a_m = \frac{2}{\pi} \int_0^\pi f(x) \sin mx \, dx \qquad (m = 0, 1, 2, \ldots)$$

$$= \frac{1}{2} \int_0^\pi \sin mx \, dx \tag{4.58}$$

$$= -\frac{1}{2m} \cos mx \Big|_0^\pi$$

where m odd $\rightarrow a_m = 1/m$
   m even $\rightarrow a_m = 0$

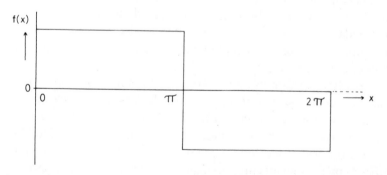

Figure 4.11. Representation of an odd square wave function.

Thus $0 < x < \pi$ yields

$$\tfrac{1}{4}\pi = \sin x + \tfrac{1}{3}\sin 3x + \tfrac{1}{5}\sin 5x + \cdots \tag{4.59}$$

(see Figure 4.12). The frequency spectrum of $F(x)$ is shown in Figure 4.13. The general description of a function by means of a summation of sine and cosine function is

$$f(x) = \int_{-\infty}^{+\infty} f(v)\, e^{2\pi i x v}\, dv \tag{4.60}$$

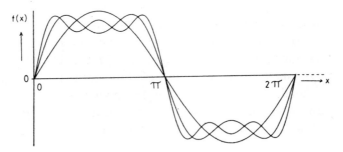

Figure 4.12. Construction of an odd square wave function by summation of sine functions (eq. 4.59).

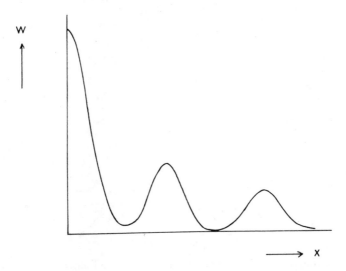

Figure 4.13. Frequency spectrum of $f(x)$ (eq. 4.60).

and of its Fourier transform

$$f(v) = \int_{-\infty}^{+\infty} f(x)\,e^{-2\pi i x v}\,dx \qquad (i = \sqrt{-1}) \qquad (4.61)$$

where $f(v)$ depends on the frequency and $f(x)$ on position or time x, or in other words $f(v)$ is defined in a frequency domain and $f(x)$ is the position or time domain. The conversion of the general equation to the sum of sine and cosine function is made by

$$e^{-2\pi i x v} = \cos(2\pi x v) - i\sin(2\pi x v) \qquad (4.62)$$

The conversion of $f(x)$ into $f(v)$ is a rather time-consuming process that can be considerably simplified by application of a fast Fourier transform algorithm and a digital computer, provided that the number of sample points in the series equals 2^m. In order to improve the S/N ratio of a signal, Fourier analysis can be applied, followed by convolution with an appropriate smoothing function, for example, an exponential function or a cutoff filter (Figure 4.14).

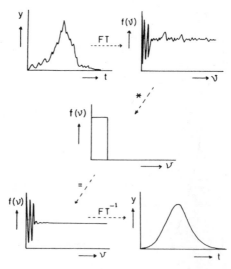

Figure 4.14. Improvement of S/N ratio by application of a Fourier transform (FT) and a cutoff filter.

A smoothing filter acts to block high frequencies and allow only low-frequency signals to pass. Some examples of smoothing filters are a triangular function ($x = 0$, $y = 1$ and $x \geqslant a$, $y = 0$); the half of a Gaussian function ($x = 0$, $y = y_{max}$ and $x = a$, $y \to 0$); and a decreasing exponential function. The selection of an appropriate smoothing function is a matter of trial and error. Sometimes the characteristics of the instrument under consideration are measured under high signal-to-noise conditions. These characteristics are used afterward in a convolution of signals measured under conditions with high noise levels. Here the spectrum including noise is convoluted by means of a Fourier transform with a virtually noise-free signal.

In the frequency domain convolution means multiplication; the place or time domain asks for a much more elaborate process for convolution. Here the convolution integral means

$$T(y) = \int_{-\infty}^{+\infty} t(x) B(y - x) dx \qquad (4.63)$$

where $T(y)$ represents a virtually noise-free convoluted signal: $t(x)$, the observed signal: and $B(y - x)$, the smoothing function. A disadvantage of convolution should be mentioned here: together with the noise part of the signal, a distortion of the signal occurs which manifests itself as a broadening of the peaks (and a lowering of the maximal signal values because of normalization).

4.3.2.4. Deconvolution

Deconvolution offers a method of eliminating distortions of signals that are caused by the instrument. Here one aims for the recovery of the original signal, for example, by application of Fourier transform analysis. As an example the distortion of an optical signal caused by the slit of the instrument is considered. The shape of the distorted signal is known to be a broadened line, and in order to recover the original line shape a transformation is required that performs the opposite of a convolution calculation:

$$T(y) = \int_{-\infty}^{+\infty} t(x) B(y - x) dx \qquad (4.64)$$

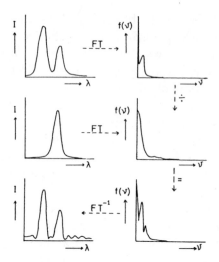

Figure 4.15. Removal of signal broadening by deconvolution of the Fourier transform (FT) of the distorted signal by the FT of the distorting signal followed by the reverse transformation.

where $T(y)$ represents the broadened signal, distorted by the Fourier transform (FT) of the slit width function $B(y - x)$; $t(x)$ equals the undistorted signal one wants to recover. The aim of the deconvolution process is the recovery of $t(x)$. In the frequency domain deconvolution means division of the FT of the observed signal by the FT of the distorting function and backtransformation of the resulting FT of the undistorted signal (HA 74a).

The deconvolution process is depicted schematically in Figure 4.15. The deconvolution results in a smaller line width and thus allows a better peak separation. However, the deconvoluted signal shows some little extra signals that are not found in the original spectrum. These signals are artificial and hamper the detection of low-intensity signals that actually might be present in the original signal.

In situations where the slit function (or other artificial disturbances due to the instrument) is unknown it should be measured separately. A general procedure accepted in spectrometry is to proceed as fillows: A spectrum of a component of a mixture is recorded with a narrow slit width A and another recording is made with identical parameter settings except for a large slit width B. It is to be noted that recording A gives a better peak separation than recording B. Now the Fourier-transformed spectrum B is divided by the FT of spectrum A in order to produce the FT of the slit function C'. The back-transformation of C' yields the slit function in the λ domain and may be used for transformation purposes in future spectra. It should be noted that this function is apparatus bound and cannot be used for transformation of spectra produced by other instruments.

4.4. DATA RECONSTRUCTION AND CURVE FITTING

The production of measurements by one-dimensional methods of analysis such as melting points, boiling points, indexes of refraction, and so on mostly do not present problems. A multiplication or division usually suffices for reconstruction of relevant information. The situation becomes more complicated for two-dimensional methods of analysis such as are encountered in spectral analysis. The analysis here means measurement of peak positions and of the heights or peak surface. Difficulties may be caused by the following:

- *Noise and drift on the original spectra.* Noise may be eliminated by smoothing. Systematic deviations in the abscissa and ordinate (the λ and the intensity scale, respectively) may be corrected by the application of calibration data, stored in advance. Baseline detection eliminates drift and the disturbances caused by slit functions can be corrected by deconvolution.

- *Overlapping peaks.* Positions of peaks can be found by differentiating the original spectrum as a function of time or wavelength. A stronger overlap asks for more consecutive differentiations in order to establish peak position. However, in practice the number of differentiations is limited because each differentiation enhances the noise. The peak surface is found by integration over time or wavelength. In order to obtain accurate peak surfaces the baseline should be well defined.

- *A baseline not sufficiently well known.* The baseline is found from those parts of a spectrum where no peaks are found. A baseline correction is only possible when such parts occur.

Another technique for separation of overlapping signals is a multiband resolution by means of curve fitting. Here it is known beforehand that the measured data follow a mathematical relationship

$$y = f(x) \tag{4.65}$$

and overlapping signals are additive. This implies for n signals

$$y(n) = \sum_{i=1}^{n} f_i(x) \tag{4.66}$$

The functions f_i are defined by a number of parameters such as peak position on the wavelength scale, peak intensity on the y-scale, half width at half height of the peak, intensity relations between groups of peaks, and so on. Each peak is described by its own function. In spectroscopy symmetrical peaks are mostly described by the Lorentz function (FR 69),

$$Y_L = \frac{A_L}{1 + B_L^2 (x - x_{max})^2} \tag{4.67}$$

by the Gauss function,

$$Y_G = A_G \exp[- B_G^2 (x - x_{max})^2] \tag{4.68}$$

or by linear combinations of these functions. Here A_L and A_G are found from intensity and/or normalization conditions such as

$$\int_{-\infty}^{+\infty} y \, dx = 1 \tag{4.69}$$

and B represents a relationship with width at half height W: $B_L = 2/W$; $B_G = 2\sqrt{(\ln 2)}/W$.

Nonsymmetric peaks are approximated by means of empirically adapted standard digital functions. The estimated line shape can be approximated as accurately as desired by enhancement of the number of points in the digital function.

4.4.1. Curve Fitting

Curve fitting aims for a mathematical description of a series of measured data by a model with adaptable parameters. In general, the following procedure is applied: A mathematical model is chosen on the basis of empirical knowledge on the system under investigation (e.g., Lorentzian line shape or Gaussian line shape). The adaptable parameters in the model such as line width, peak position, and peak height are calculated in a manner that makes the sum of the squares of the deviations between calculated and measured data points as small as possible (SE 79, VA 75). The number of parameters to be found equals the number of peaks multiplied by the number of parameters per peak. Suppose the analytical function that is going to represent the measured data equals

$$y = f(\alpha_1, \alpha_2, \alpha_3, \ldots, \alpha_n, x) \qquad (4.70)$$

where $\alpha_1, \ldots, \alpha_n$ are parameters to be determined such as line width and peak position. It is assumed that the number of measured data m is greater than the number of independent parameters n. The residual sum of squares R that represents the differences between the calculated values and the measured data Y_i is defined by

$$R = \sum_{i=1}^{m} [y_i - f(\alpha_1, \alpha_2, \ldots, \alpha_n, x)]^2 \qquad (4.71)$$

and has to be minimized with respect to each α_i. The condition

$$\frac{\partial R}{\partial \alpha_1} = \frac{\partial R}{\partial \alpha_2} \cdots \frac{\partial R}{\partial \alpha_n} = 0 \qquad (4.72)$$

yields n equations of the form

$$-2[y_1 - f(\alpha_1, \alpha_2, \ldots, \alpha_n, x_i)]\frac{\partial f_i}{\partial \alpha_j} = 0 \quad (j = 1, 2, \ldots, n) \qquad (4.73)$$

In eq. 4.71 the residual sum of squares R is defined on an absolute basis. It is sometimes desirable, however, to fit the data on a percentage basis. This is particularly true when the magnitudes of the α_j values are spread over a wide range. Using W_i as a weighing factor, we can redefine R as

$$R = \sum_{i=1}^{m} W_i [y_i - f(\alpha_1, \alpha_2, \ldots, \alpha_n, x_i)]^2 \qquad (4.74)$$

The iterative least-squares method of problem solution is applied to situations where a number of equations (data) is available that exceeds the number of unknown parameters; for example, when a UV absorption spectrum of a mixture of two components is given as well as the individual spectrum of each of the components and the concentration ratio in the mixture is unknown, two extinction measurements would suffice to produce an answer. However, inclusion of all extinction measurements and a least-squares calculation enhances the accuracy and reliability of the answer.

Properties of curve fitting by least-squares procedures may be summarized as follows:

- All data must be stored in a data set before the calculations can be started.
- The baseline adaptation should be taken into account.
- Partial derivatives with respect to all adaptable parameters should be calculated at each position of the spectrum. Sometimes this calculation must be repeated at each iteration cycle.
- Matrix manipulations are necessary.
- The starting parameters should be estimated. The possible dependence of the solution on the starting parameters must always be tested.
- Convergence may be accelerated by properly diminishing the parameter variations.
- The whole procedure may be a failure if an inappropriate mathematical model is selected.

In spectrum simulations the following limitations are set:

- The mathematical model for the line shape should be correct.
- Eventually a standardized digital line shape function may be applied; however, all signals must have the same line shape.

In situations where a spectrum contains Lorentzian and Gaussian lines it is not possible to simulate the entire spectrum with one line shape function.

In chromatographic spectra it is sometimes seen that high-retention signals "tail" more than low-retention signals. Here, too, one single line shape function cannot simulate all signals. Hence—

- The number of signals must be known before the iterative least-squares method is started. This knowledge may result from a theoretical consideration or from a test procedure that correlates the number of peaks in the model with the reduction of the average standard deviation between model and measurements.
- Even situations where the above conditions are fulfilled may not be uniquely solvable because of serious interaction between two or more of the parameters. Such situations are met with overlapping signals.

Another approach is the use of Kalman filters (Section 4.3.2.2), treated by Poulisse (PO 79), Brown (BR 86), and Rutan (RU 89).

4.5. DATA HANDLING

In an analytical laboratory sometimes the relation between two measured quantities on a subject or a series of subjects is investigated. Two important questions arise: (1) what is the accuracy of the measurements; and, related to this, (2) how accurate is the relation between the measured quantities? The problems related to these questions may be attacked by means of statistical techniques such as autocorrelation techniques and covariance calculations.

4.5.1. Mean, Variance, and Covariance

In order to do an accurate measurement, it can be repeated a number of times, keeping the conditions of the measuring system as constant as possible. The result of such a series of measurements should be independent of time when the measuring system and the measuring quantities pertain to stationary states. In statistics the average result of such a series of measurements is sometimes referred to as expectation value $E(x)$ or expectation of x:

$$E(x) = \int_{+\infty}^{-\infty} x f(x)\, dx \qquad (4.75)$$

or

$$E(x) = \sum_{-\infty}^{+\infty} x p(x) \tag{4.76}$$

Equation 4.75 refers to a situation where x is described by a continuous probability distribution function $f(x)$, whereas eq. 4.76 is applied for discrete values of x occurring with a probability $p(x)$.

The definition of the expectation value of x can be extended to the expectation value of a function of x, $g(x)$ being a function of which x has an expectation value

$$E[g(x)] = \sum_{-\infty}^{+\infty} g(x) p(x) \tag{4.77}$$

for discrete x values and

$$E[g(x)] = \int_{-\infty}^{+\infty} g(x) f(x) \, dx \tag{4.78}$$

for continuous x values. For $g(x) = ax$, a representing a constant, this results in

$$E[g(x)] = E[(ax)] = \sum_{-\infty}^{+\infty} ax\, p(x) = a \sum_{-\infty}^{+\infty} x p(x)$$
$$= aE[x] \tag{4.79}$$

Equation 4.79 also holds for the continuous probability distribution function.)

Apart from the expectation value or mean value of x one may also be interested in the spread of x values over the series of measurement in order to gain insight into the reproducibility of the measurements. The statistical expression applicable here is the variance of x,

$$\mathrm{var}(x) = E\{[x - E(x)]^2\}$$
$$= \sum_{-\infty}^{+\infty} [x - E(x)]^2 p(x) \tag{4.80}$$

or in the situation with a continuous probability distribution of x,

$$\text{var}(x) = \int_{-\infty}^{+\infty} f(x)[x - E(x)]^2 \, dx \qquad (4.81)$$

Another expression used is the standard deviation $\sigma(x)$:

$$\sigma_x = [\text{var}(x)]^{1/2} \qquad (4.82)$$

again selecting the situation $E(ax)$ one may derive

$$\text{var}(ax) = a^2 \text{var}(x) \qquad (4.83)$$

and

$$\sigma(ax) = |a| \sigma_x \qquad (4.84)$$

where $|a|$ is the absolute value of a.

Other useful equations that are easily derived are for $z = x + y$:

$$\text{var}(z) = \text{var}(x) + \text{var}(y) + 2 \text{cov}(x, y) \qquad (4.85)$$

Here the covariance between x and y is called $\text{cov}(x, y)$

$$\text{cov}(x, y) = E\{[x - E(x)][y - E(y)]\} \qquad (4.86)$$

In situations where the x values are represented by a Gaussian probability distribution, $E(x)$ gives the position of the maximum; $\text{var}(x)$ is related to the spread of x around the most probable value; $\text{cov}(x, y)$ is related to the relationship between x and y values. (Its meaning can be inferred from so-called contour diagrams.) In Figure 4.16a high values of x are correlated with low values of y. This situation is described by a high negative covariance between x and y. In Figure 4.16b high x values correspond with high y values; here the $\text{cov}(x, y)$ is high and positive. In Figure 4.16c there is no correlation between x and y; thus $\text{cov}(x, y) = 0$. Figure 4.16d depicts a situation with a positive $\text{cov}(x, y)$; however, its magnitude is considerably lower than that of Figure 4.16b.

The magnitude of the covariance can be standardized. This standardized covariance is called the correlation coefficient $\rho(x, y)$:

$$\rho(x, y) = \frac{\text{cov}(x, y)}{\sigma(x)\sigma(y)} \qquad (4.87)$$

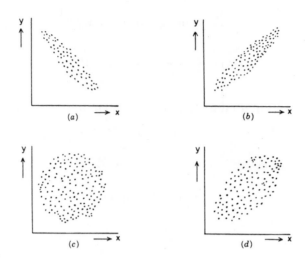

Figure 4.16. Correlation between x and y values: (a) negative correlation with low spread; (b) positive correlation with low spread; (c) no significant correlation; (d) positive correlation with high spread (low significance).

in which $\sigma(x) = [\text{var}(x)]^{1/2}$. A special situation is met when $y = ax$ $(a \geqslant 0)$. Here

$$\text{var}(x) = a^2 \text{var}(x) \tag{4.88}$$

$$\begin{aligned}
\text{cov}(x, y) &= E\{[x - E(x)][y - E(y)]\} \\
&= E\{[x - E(x)][ax - aE(x)]\} \\
&= a\,\text{var}(x)
\end{aligned} \tag{4.89}$$

Thus

$$\rho(x, y) = \frac{a\,\text{var}(x)}{\sigma(x)\sigma(y)} = \frac{a\,\text{var}(x)}{\sigma(x)\,a\,\sigma(x)} = 1 \qquad (a \geqslant 0) \tag{4.90}$$

In situations where the value found for x fixes completely the value to be expected for y, $\rho(x, y) = \pm 1$. Where no relationship exists $\text{cov}(x, y) = 0$ and thus $\rho(x, y) = 0$.

In the situations considered so far x as well as y pertain to stationary states. This changes when time-dependent processes are considered. Here the measured values x vary with time. The collected values x over a period of time are called a time series.

4.5.2. Time Series

A time series is a collection of measurements on the same variable quantity, measured as a function of time. Continuous time series are found in situations where the variable is continuously measured and recorded (e.g., temperature, pressure). In analytical chemical laboratories many types of measurements are performed at discrete time intervals that produce discrete time series. The conversion of a continuous time series into a discrete one is achieved only by two methods:

1. By sampling of a continuous time series
2. By accumulation over a given period of time (e.g., collection of rainwater)

From here on only discrete time series are discussed. In order to discuss the behavior of a time series the following notation is used: observations are made at times $T_0, T_1, T_2, \ldots, T_n$, giving observed values $Z(T_0), Z(T_1), \ldots, Z(T_n)$. In situations where the time interval between two successive measurements is a constant h the notation may conveniently be changed into

measuring time $T_0, T_0 + h, T_0 + 2h, \ldots, T_0 + nh$

measurement $Z_0, Z_1, Z_2, \ldots, Z_n$

For a starting time $T_0 = 0$ and h equal to the unit of time, Z_t becomes the measurement at time t. In practice two types of time series may be distinguished (BO 70):

- *Deterministic time series* — in which the future is completely determined, such as

$$Z_t = \cos(2\pi f t) \qquad (4.91)$$

- *Statistical time series* — in which the values expected may be forecast in terms of a probability distribution function

A statistical phenomenon that develops in time according to probability laws is called a *stochastic process* (Figure 4.17). In order to

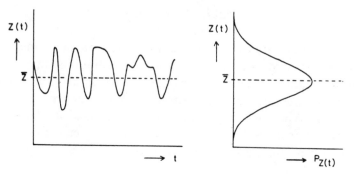

Figure 4.17. Example of a stochastic process $Z(t)$ as a function of time with its probability distribution function $P_{z(t)}$.

analyze, that is, describe, a statistical time series, it is viewed as being a particular realization of a stochastic process.

A stationary process is a special type of stochastic process: it is a process in statistical equilibrium, which means that the properties of the process do not depend on the starting time of the observation period. Such a process can be subdivided into parts so that all may start at time T_0. This subdivision is allowed, provided that a minimal realization length is obtained or surpassed. Such signals with equal statistical properties are called *ergodic*. Because the process is statistically determined, each time t provides a probability distribution for Z. This implies that a time series of M observations should be represented by an M-dimensional vector $(Z_1, Z_2, \ldots, Z_m)$ with a probability distribution $P(Z_1, Z_2, \ldots, Z_m)$. Given a stationary process, the condition holds that m measurements $Z_{t1}, Z_{t2}, \ldots, Z_m$ measured at time $t_1, t_2, \ldots, t_m$ are the same as the measurements $Z_{t1+v}, Z_{t2+v} \ldots, Z_{tm+\tau}$ measured at $t_{1+v}, t_{2+v}, \ldots$. Two important properties of stationary processes are the following:

- The mean value μ is a constant given by

$$\mu = E[Z_t] = \int_{-\infty}^{+\infty} Z P(z) \, dz \tag{4.92}$$

- The variance of Z is a constant

$$\sigma_z^2 = E[Z_t - \mu)^2] = \int_{-\infty}^{+\infty} (z - \mu)^2 \, P(z) \, dz \tag{4.93}$$

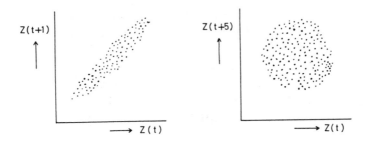

Figure 4.18. Decrease in correlation with increasing time span of a time series $Z(t)$.

In the special situation $m = 1$ the probability distribution $P(Z_t)$ is the same for all t values and can be denoted as $P(z)$.

Here $P(z)$ can be found from a histogram of measurements on time series $Z_1, Z_2, Z_3, \ldots, Z_n$ because $P(z)$ is a constant for all values of t. The average value μ from a stationary process is found from the average value $\bar{Z}$ of the time series of $n \to \infty$

$$\bar{Z} = \frac{1}{n} \sum_{t=1}^{n} Z_t \tag{4.94}$$

and the variance

$$\sigma^2 = \frac{1}{n-1} \sum_{t=1}^{n} (Z_t - \bar{Z})^2 \tag{4.95}$$

(for large values of n, $n - 1$ approaches n).

For a stationary process the probability of occurrence of a pair of measurements $P(Z_{t1}, Z_{t2})$ is a constant for all times t_1 and t_2 provided that the time interval between t_1 and t_2 is a constant. This is illustrated in Figure 4.18, where it is apparent that the correlation between Z_{t+1} and Z_t is higher than the between Z_{t+5} and Z_t. In general the correlation decreases upon an increase of τ.

4.5.2.1. Autocorrelation

The covariance between X_t and $Z_{t+\tau}$ of a time series can be calculated and is called the autocovariance γ_τ (*auto* because one stays within the time series):

$$\gamma_\tau = \text{cov}[Z_t, Z_{t+\tau}] \tag{4.96}$$
$$= E[(Z_t - \mu)(Z_{t+\tau} - \mu)]$$

The collection of γ values for various values of π normalized to the variance is called the autocorrelation function ρ_τ:

$$\sigma_\tau = \frac{E\left[(Z_t - \mu)(Z_{t+\tau} - \mu)\right]}{\{E\left[(Z_t - \mu)^2\right] E\left[(Z_{t+\tau} - \mu)^2\right]\}^{1/2}}$$

$$= \frac{E\left[(Z_t - \mu)(Z_{t+\tau} - \mu)\right]}{\sigma_z^2} \tag{4.97}$$

Because $E[(Z_t - \mu)^2 = E[(Z_{t+\tau} - \mu)^2] = \gamma_0$ for a stationary process,

$$\rho_t = \frac{\gamma_\tau}{\gamma_0} \tag{4.98}$$

implying

$$\rho_0 = \frac{\gamma_0}{\gamma_0} = 1 \tag{4.99}$$

Another symbol for γ_τ, the autocovariance function, is $\psi_{zz}(\tau)$, and another symbol for ρ_t, the autocorrelation function, is $\varphi_{zz}(\tau)$; the latter function has no dimension. Thus,

$$\gamma_\tau = \rho_\tau \sigma_z^2 \tag{4.100}$$

and

$$\psi_{zz}(\tau) = \varphi_{zz}(\tau)\tau_z^2 \tag{4.101}$$

Because τ is between $-\infty$ and $+\infty$ and the function is symmetric around $\tau = 0$, in general only that part of the function between 0 and $+\infty$ is represented. The time series given in Figure 4.19 yields the autocorrelation function. The autocorrelation function for $\tau = 0$ equals 1.0; a higher value of τ results in a lower value of $\varphi_{zz}(\tau)$. For the noncorrelated measurements the autocorrelation function sharply drops to 0 for all $\tau > 0$. Such an autocorrelation function also results from a series of random numbers, that is, white noise. In practical situations correlated time series frequently produce an autocorrelation function that may be represented by an exponential function:

$$\varphi_{zz}(\tau) = \exp(-\tau/T_x) \tag{4.102}$$

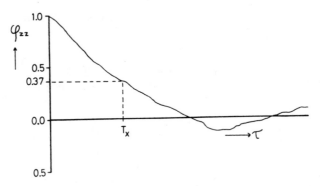

Figure 4.19. Autocorrelation function φ_{zz} of a time series $Z(t)$ as a function of time.

Here T_x is called the time constant of the process. It is found from the τ value at $\varphi_{zz}(\tau) = 0.37$, because $\ln 0.37 = -1 = -\tau/T_x$.

A nonstationary process yields $\varphi_{zz}(\tau)$ values that do not tend to 0 for high τ values. A measured signal that includes drift is always autocorrelated and produces a constant value for $\varphi_{zz}(\tau)$. A periodic function such as a sine function yields an autocorrelation function with the same periodicity and shape. A series of measurements suffering from many random high-frequency disturbances can produce a very satisfactory autocorrelation function. The calculation of $\varphi_{zz}(\tau)$ may be useful in situations where a series of process values is hidden by sources of noise. Because white noise produces a zero contribution to the autocorrelation function for $\tau \neq 0$, the influence of random disturbances on the autocorrelation function is eliminated and the calculation of $\varphi_{zz}(\tau)$ offers a method of eliminating noise from a time series of measurements.

A stationary time series is characterized by the mean value μ and the autocovariance $\psi_{zz}(\tau)$ or alternatively by μ, the variance σ_z^2, and $\varphi_{zz}(\tau)$, the autocorrelation function.

A signal containing sufficient information for a process description is called an ergodic signal. The autocorrelation function of a stationary process has a limit-value zero for $\tau \to \infty$:

$$\lim_{\tau \to \infty} \varphi_{zz}(\tau) = 0 \qquad (4.103)$$

and thus a signal with a zero-limiting autocorrelation function for $\tau \to \infty$ pertains to a stationary process and is an ergodic signal. This implies that a finite time series of measurements suffices for the process description.

Calculation of Autocorrelation Functions. For a given time series $t_1, t_2, \ldots, t_n$ with $n \gg \tau$, $n - \tau \approx n$. The autocovariance equals

$$\psi_{zz}(\tau) = \frac{1}{n - \tau} \sum_{t=1}^{n-\tau} (Z_t - \bar{Z}_t)(Z_{t+\tau} - \bar{Z}_{t+\tau}) \tag{4.104}$$

For a stationary process

$$\bar{Z}_t = \bar{Z}_{r+\tau} = \bar{Z} \tag{4.105}$$

and thus

$$\psi_{zz}(\tau) = \frac{1}{n - \tau} \sum_{t=1}^{n-\tau} (Z_t Z_{t-\tau} + \bar{Z}^2 - Z_t \bar{Z} - Z_{t+\tau} \bar{Z})$$

$$= \frac{1}{n - \tau} \sum_{t=1}^{n-\tau} (Z_t Z_{t+\tau}) - \bar{Z}^2 \tag{4.106}$$

The variance equals $\psi_{zz}(\tau)$ for $\tau = 0$:

$$\sigma_z^2 = \frac{1}{n} \sum_{t=1}^{n} (Z_t)^2 - \left(\frac{1}{n} \sum_{t=1}^{n} Z_t \right)^2 \tag{4.107}$$

and the autocorrelation function

$$\varphi_{zz}(\tau) = \frac{\psi_{zz}(\tau)}{\sigma_z^2} \tag{4.108}$$

Autocorrelation Function as a Predicting Function. In most practical situations the autocorrelation function $\varphi_{zz}(\tau)$ of a time series $Z(\tau)$ equals

$$\varphi_{zz}(\tau) = \exp(-\tau/T_x) \tag{4.109}$$

This may be used to predict a process value for time t, based on the process value Z_0 at time 0 (MU 78a):

$$Z_t = Z_0 \exp(-\tau/T_x) \tag{4.110}$$

where Z_t = predicted process value for time t

Z_0 = measured process value at time 0

τ = time span for the prediction

T_x = time constant of the process $Z(\tau)$

The accuracy of the prediction Z_t decreases with an increased τ value. With $\tau \to \infty$, a prediction of all values is possible within the limit of standard deviation and a probability corresponding to the Gaussian distribution.

Thus the autocorrelation function yields information on the transient behavior of a time series.

4.5.2.2. Cross-Correlation

The autocorrelation function correlates a signal, for example, a series of measurements on a process, with itself. Another possibility arises when two different time series are intercorrelated, for instance, the input time series $x(t)$ and the output series $y(t)$ of a process at a time τ after the input measurement (Figure 4.20).

The autocovariance of $x(t)$, ψ_{xx}, and the cross-covariance function of $x(t)$ and $y(t)$, $\psi_{x,y}$, can be represented as shown in Figure 4.21. The mathematical expression for the cross-covariance function x with y

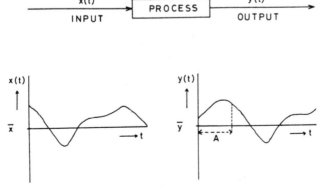

Figure 4.20. Input and output signals of a process. The output signal $y(t)$ has a delay A with respect to the input signal $x(t)$.

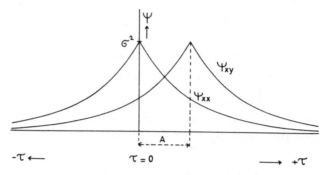

Figure 4.21. Autocovariance function $\psi_{x,x}$ and cross-covariance function $\psi_{x,y}$ of the signals from Figure 4.20. Here σ^2 represents the variance of the input signal $x(t)$; A is the delay between input $x(t)$ and output $y(t)$.

reads

$$\psi_{x,y}(\tau) = E\left(\{x(t) - E[x(t)]\}\,\{y(t+\tau) - E[y(t+\tau)]\}\right) \quad (4.111)$$

Here we see that $\psi_{x,y}(\tau)$ for $\tau = 0$ is smaller than $\psi_{x,x}(\tau)$ for $\tau = 0$. An increase in τ results in an increased value for $\psi_{x,y}(\tau)$ until a maximal value is obtained; afterward the value of $\psi_{x,y}(\tau)$ decreases and tends toward 0. It is clear that there is a time lag between input signal $x(t)$ and output signal $y(t)$. In contrast to the autocovariance function, the cross-covariance function is mostly not symmetric with respect to $\tau = 0$. The position of the maximum cross-covariance is given by the time lag between $x(t)$ and $y(t)$:

A positive time lag means $x(t)$ is advanced with respect to $y(t)$.
A negative time lag means $x(t)$ is retarded with respect to $y(t)$.
A zero time lag means $x(t)$ and $y(t)$ are in phase.

Another property of $\psi_{x,y}(\tau)$ reads

$$\psi_{x,y}(\tau) \leqslant [\psi_{xx}(0)\cdot\psi_{yy}(0)]^{1/2} \quad (4.112)$$

The covariance of the sum of independent signals equals the sum of the covariance of the individual signals:

$$\psi_{x+y}(\tau) = \psi_x(\tau) + \psi_y(\tau) \quad (4.113)$$

The importance of the cross-correlation function $\rho_{x,y}(\tau) = \psi_{x,y}/(\sigma_x \sigma_y)$ is in its ability to describe a process. An input signal $x(t)$, corresponding with an autocorrelation function $\varphi_{xx}(\tau)$, determines an output signal $y(t)$ found from $\varphi_{x,y}(\tau)$.

Another procedure for identifying a process is the impulse response method. Here the response of a process to a well-known input pulse is characterized. An advantage of correlation functions over impulse response measurements is that, as well as autocorrelation, cross-correlation can be calculated from measurements available from the history of a process. In order to apply impulse response methods, the normal behavior of a process has to be disturbed. Apart from this it is sometimes rather difficult to produce a pulse disturbance on the input signal of a process.

4.6. ANALYSIS OF VARIANCE

The analysis of variance is a statistical technique for subdividing the total variance of a set of observations into partial variances that are attributed to well-defined sources. For example, when the quality of a basic material is given by the quantity of a certain chemical compound in the material, one may measure the quality by analyzing a sample of the material. The analytical data obtained from various samples of the same material generally are distributed around an average value.

This distribution of results may be caused by the method of sampling, but also by the method of analysis. The aim of the analysis of variance is to attribute part of the variance to the method of sampling and another part to the method of analysis. An eventual improvement of the measurement of the quality of the material may result from a decrease in the variance of the distribution in the measurements. Here the cause of the most important source of variance can be found and its influence can be diminished.

4.6.1. Definitions

The variance s^2 of a set of n measured values x_i equals the arithmetic mean of the quadratic deviation of the measurements with respect to their average value $\bar{x}$:

$$s^2 = \frac{1}{n-1} \sum_{i=1}^{n} (x_i - \bar{x})^2 \qquad (4.114)$$

with

$$\bar{x} = \frac{1}{n} \sum_{i=1}^{n} x_i \qquad (4.115)$$

In eq. 4.115 n, the number of measurements in the set, is limited. The same equations are used for the variance of an idealized set of measurements, that is, a set where the population is increased to such an extent that the application of statistical theories is considered to be correct. The variance of such a set is called σ^2; the average value of the measurements, μ; and the population, M ($M \gg n$). In contrast to $\bar{x}$, μ is called the "true" mean value. Because of its magnitude, $M - 1$ (the value of the denominator in eq. 4.114) can be converted into M.

4.6.1.1. Degrees of Freedom

When out of a set of three observations two observations are given, together with the average value of the set, the value of the third observation is fixed; for example, $x_1 = 2$, $x_2 = 4$, $\bar{x} = 5$ yields $x_3 = 9$. The number of degrees of freedom of this example is 2. Generally the number of degrees of freedom equals the number of independent data from a series of observations, or, in other words, the total number of data diminished by the number of independent restrictions.

4.6.2. Conditions on Applicability

In the method of analysis of variance, a considerable number of observations are involved, collected according to a previously designed scheme. The observations are divided into groups resulting from various samples, all belonging to a given population. In order to compare the results from different samples a number of conditions are set:

1. The population investigated should be at least approximately normally distributed, in order to allow calculation methods based upon the variances and the application of the F-test.

2. All samples must be taken from the same population. Comparison of data resulting from different methods of analysis or different instruments in order to investigate the effect of another method or instrument of course involves data from different populations. Here the method of comparing variances looks the same but the conclusions to be drawn are quite different, namely, the preference of one method or instrument over another.

3. The population on which the measurements are made should be constant in time. The measurements are made consecutively; therefore a time dependence of the population may result in a trend in the collected data and thus a wrong estimate of the variance can be the consequence.

4. The sampling of the probes under investigation should be such that they are normally distributed because here too calculations of variances and F-tests are performed.

5. All probes should have the same variance. This condition is crucial for the analysis of variance method. The equal variance in combination with the equal mean value of all probes is called the null hypothesis of the analysis, and—based on the F-tests—the acceptance or rejection of this hypothesis provides the conclusion about the unequal contribution of variances by various sources. The design of a proper analysis of variance scheme employs at least two different methods for estimation of the common variance of the total population. When two such methods provide different estimates of the mean value, and these differences are proved to be significant, it is concluded that the averages of the samples are different. The null hypothesis is rejected.

The method is illustrated by the following example. From a given population of objects k samples are selected, each containing n_k objects. The n_k objects of sample k are normally distributed with a variance σ_k^2 that equals σ_0^2. Here σ_0^2 is the variance of the parent population. The question to be asked is whether the mean values $\bar{x}_k$ of the samples k are equal or not. This question is formulated in the null hypothesis H_0:

$$\mu = \mu_0 \qquad (i = 1, 2, 3, ..., k) \qquad (4.116)$$

The analysis of variance scheme designed to answer this question can be summarized as follows. Calculate for each sample j:

1. The number of observations made on sample j equals n_j; the observations themselves are $x_{i,j}$ ($i = 1, 2, ..., n_j$).
2. The average value of x found for sample j equals

$$\bar{x}_j = \frac{1}{n_j} = \sum_{i=1}^{n_j} x_{i,j} \tag{4.117}$$

3. The variance of the observations made on sample j equals

$$s_j^2 = \frac{1}{n_j - 1} \sum_{i=1}^{n_j} (x_{i,j} - \bar{x}_j)^2 \tag{4.118}$$

Thus the k calculated values s_j^2 all are estimates of the population variance σ_0^2 of the parent distribution; this is a condition set in advance in order to apply the analysis of variance method. This being so, all s_j^2 values may be combined in order to yield a better value for σ_0^2. This value equals

$$s_w^2 = \frac{1}{k} \frac{\sum_{j=1}^{k} \sum_{i=1}^{n_j} (x_{i,j} - \bar{x}_j)^2}{n_j - 1} \tag{4.119}$$

This better estimate of σ_0^2 is based on the variance estimates within the sample, and is called the within-group variance.

Now another way is designed to obtain an estimate of σ_0^2, starting from the mean values $\bar{x}_j$ of all samples k. Suppose these $\bar{x}_j$ values stem from a population with variance σ_0^2. The variance of the $\bar{x}_j$ values equals

$$s_x^2 = \frac{1}{n_j} \sigma_0^2 \tag{4.120}$$

This variance is also found from

$$s_x^2 = \frac{1}{k-1} \sum_{j=1}^{k} \frac{n_j}{n} (\bar{x}_j - \bar{x})^2 \tag{4.121}$$

where

$$\bar{n} = \sum_{j=1}^{k} (n_j/k) \tag{4.122}$$

and

$$\bar{x} = \frac{\sum_{j=1}^{k} \sum_{i=1}^{n_j} x_{i,j}}{\sum_{j=1}^{k} n_j} \tag{4.123}$$

When it is assumed (eq. 4.121) that all samples k contain the same number of measurements, then $n_j = n$ for all values of j. For unequal values of n_j the terms $(\bar{x}_i - \bar{x})^2$ should be multiplied by a proper weighting factor $n_j/\bar{n}$ in order to equalize the contribution of all measurements $x_{i,j}$ to the variance.

This second estimate of σ_0^2, which is based on the variance between the mean values of the samples, is called the between-group variance S_b^2. Its value equals ns_x^2, provided all samples contain the same number of measurements n.

The null hypothesis states that s_b^2 is an estimate of σ_0^2; s_w^2 is always an estimate of σ_0^2. If s_b^2 and s_w^2 are unequal, tested for significance by the F-test, this cannot be caused by unequal values for the variance, because the application of the analysis of variance scheme implies equal values for sample variances. Therefore the mean values $\bar{x}_k$ should be unequal and the null hypothesis must be rejected. Equal values found for s_b^2 and s_w^2 imply that the null hypothesis is correct and that s_b^2 and s_w^2 may be combined in order to obtain a good estimate s_0^2 for σ_0^2:

$$s_0^2 = \frac{(n_1 - 1) s_w^2 + (n_2 - 1) s_b^2}{n_1 + n_2 - 2} \tag{4.124}$$

in which $n_1 + 1$ equals the number of degrees of freedom in s_w^2, that is, $k(n - 1)$, and $n_2 - 1$ equals the number of degrees of freedom in s_b^2, that is, $k - 1$. Thus

$$s_0^2 = \frac{\sum_{i=1}^{n} \sum_{j=1}^{k} (x_{i,j} - \bar{x})^2 + n \sum_{j=1}^{k} (\bar{x}_j - \bar{x})^2}{kn - 1} \tag{4.125}$$

It is assumed that $n_j = n$ for all values $j = 1, ..., k$.

If H_0 has to be rejected, s_b^2 is not a good estimate of σ_0^2 but s_w^2 is. Here s_b^2 is interpreted as being the true variance of the parent population, increased by an amount $n\sigma_\mu^2$; σ_μ^2 equals the variance of the population of the mean values μ_i of the k samples, each with n numbers. The estimated value of σ_μ^2 equals s_μ^2 and is found from

$$s_\mu^2 = (s_b^2 - s_w^2)/n \qquad (4.126)$$

Generally $\mu^2 \geqslant 0$, so $s_b^2 \geqslant s_w^2$. If $s_b^2 < s_w^2$, the difference between the two should not be significant. If the difference *is* significant, a mistake has been made in the calculations or in the measurements.

A nonzero contribution s_μ^2 to the variance between samples is interpreted as being the introduction of uncertainty to the analytical results by the method of sampling; s_w^2 is the uncertainty in the analytical results caused by the method of analysis itself or by the distribution of x *in* the parent population.

The method can be extended to more variables by grouping all measurements in such a way that one parameter of a group is kept constant each time. The difference between the estimated variance of such a group and the best estimated value of σ_0^2 each time is tested on significance, and if this difference is significant it is attributed to uncertainty of the parameter under investigation.

Table 4.9 is an example of an analysis of variances scheme. When the measurements of Table 4.10 are used, Table 4.11 results.

Table 4.9. Analysis of Variance Scheme

Source of Variance	Sum of Squares	Number of Degrees of Freedom	Mean Square	Quantity Estimated
Between samples	$S_b = n \sum_{j=1}^{k} (\bar{x}_j - \bar{x})^2$	$\nu_1 = k - 1$	$s_b^2 = \dfrac{S_b}{k-1}$	$\sigma_0^2 + n\sigma_\mu^2$
Within samples	$S_w = \sum_{j=1}^{k} \sum_{i=1}^{n} (x_{i,j} - \bar{x}_j)^2$	$\nu_2 = k(n-1)$	$s_w^2 = \dfrac{S_w}{k(n-1)}$	σ_0^2
Total	$S_t = \sum_{j=1}^{k} \sum_{i=1}^{n} (x_{i,j} - \bar{x})^2$	$\nu_3 = nk - 1$		

Table 4.10. Numerical Example of an Analysis of Variance with Two Variables[a]

Analysis No.	Sample No.:	Analytical Data of Five Analyses on Six Samples					
		1	2	3	4	5	6
1		145	140	195	45	195	120
2		40	155	150	40	230	55
3		40	90	205	195	115	50
4		120	160	110	65	235	80
5		180	95	160	145	225	45
Average value of sample, $\bar{x}_j$		105	128	164	98	200	70

[a] The $x_{i,j}$ values, denoting the amount of a given chemical compound of a sample j, are listed. Each sample has been analyzed five times ($i = 1, 2, \ldots, 5$).

Table 4.11. Analysis of Variance Scheme, Numerical Example

Source of Variance	Sum of Squares	Number of Degrees of Freedom	Mean Square	Quantity Estimated
Between samples	$S_b = 56{,}357$	$v_1 = 5$	$s_b^2 = 11{,}272$	$\sigma_0^2 + 5\sigma_\mu^2$
Within samples	$S_w = 58{,}830$	$v_2 = 24$	$s_w^2 = 2{,}451$	σ_0^2
Total	$S_t = 115{,}187$	$v_3 = 29$		

Conclusions:

$$s_b^2 = 11{,}272, \qquad S_w^2 = 245 \rightarrow S_b^2 > S_{2w}$$

$$F_{(v_1, v_2, \alpha)} = F_{(5, 24, 0.05)} = 2.62 \qquad \text{(see Fisher Table 3.16)}$$

$$F = \frac{s_b^2}{s_w^2} = 4.50$$

$F > F_{(v_1, v_2, \alpha)} \rightarrow S_b^2$ is significantly higher than s_w^2

H_0 rejected $S_\mu^2 = \dfrac{s_b^2 - s_w^2}{5} = 1764$

S_μ^2 is an estimate of σ_μ^2, the variance of the average analyses of the samples

4.6.3. Hierarchical Classification

The example presented above can be depicted as in Figure 4.22. The total variance σ_0^2 of all analytical results equal the sum of the contributions of each level in the hierarchical one. This scheme may easily be extended to more levels, each with its own cause of errors.

4.6.4. Cross-Classification

This type of classification aims for classification of observations in a manner that is independent of all other causes of error. As an example, in comparing three different analytical procedures in five different laboratories, the variable parameters are the laboratories and the procedures. The laboratories and the procedures are completely independent, and the results of the measurements can be grouped according to laboratories and according to analytical procedures. Discrepancies between laboratories can be found from the average analytical results of the various procedures; discrepancies between procedures, from the average results of various laboratories. All data suffer from errors in the measurements or sampling that contribute to the standard deviation of the interlaboratory as well as the interprocedure σ^2. The cross-classification scheme can be extended to more than two variables.

The matrix describing the cross-classification scheme for n laboratories and m procedures is given in Table 4.12.

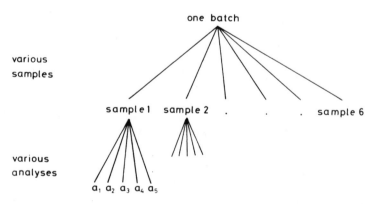

Figure 4.22 Hierarchical classification in an analysis of variance scheme.

Table 4.12. Analysis of Variance Scheme (Cross-Classification)

Laboratory	Procedure			
	1	2	$\cdots$	m
1	$x_{1,1}$	$x_{1,2}$	$\cdots$	$x_{1,m}$
2	$x_{2,1}$	$\cdots$	$\cdots$	$\cdots$
3	$\cdots$	$\cdots$	$x_{r,c}$	$\cdots$
4	$\cdots$	$\cdots$	$\cdots$	$\cdots$
$\vdots$	$\vdots$	$\vdots$	$\vdots$	$\vdots$
n	$x_{n,1}$	$\cdots$	$\cdots$	$x_{n,m}$

The analysis of variance scheme is given in Table 4.13. In this scheme $D \neq A + B$ for most situations. This may be caused by analysis errors and mutual interaction effects between laboratories and procedures that are not taken into account. The value of $C = D - (A + B)$ is called the rest variance. When no interaction effects are expected between the two parameters, the value of C equals an estimate of σ_0^2.

4.7. PRINCIPAL COMPONENT ANALYSIS

4.7.1. Introduction

In analytical chemistry it is often the case that several parameters are available per sample. Imagine that a lake suspected of being polluted is sampled in order to find out the quality of the water system. On these samples several analyses are carried out such as those of heavy metals and of organic compounds, including polychlorobiphenyls (PCBs) and polycylic aromatic hydrocarbons (PAHs); sample location and other relevant information is also available. This results in a data table like that shown in Table 4.14.

Mainly owing to the integration of computers and measurement techniques, many other examples can be found in analytical chemistry where a large amount of data can be obtained per sample in a short time period. Inductively coupled plasma spectroscopy, for example, enables several elements to be measured simultaneously. With capillary GC it is common practice to analyze 50–100 components in one run. Another example is process control where many process parameters

Table 4.13. Analysis of Variance Scheme for Cross-Classification[a]

Cause of Variation	Sum of Squares	Degrees of Freedom	Mean Square	Quantity Estimated
Between rows (laboratories)	$A = m \sum\limits_{i=1}^{n} (\bar{x}_i - \bar{x})^2$	$v_1 = n - 1$	A/v_1	$\sigma_0^2 + m\sigma_n^2$
Between columns (procedure)	$B = n \sum\limits_{j=1}^{m} (\bar{x}_j - \bar{x})^2$	$v_2 = m - 1$	B/v_2	$\sigma_0^2 + n\sigma_m^2$
Error	$C = \sum\limits_{i=1}^{n} \sum\limits_{j=1}^{m} (x_{i,j} - \bar{x}_i - \bar{x}_j + \bar{x})^2$	$v_3 = (m-1)(n-1)$	C/v_3	σ_0^2
Total	$D = \sum\limits_{i=1}^{n} \sum\limits_{j=1}^{m} (x_{i,j} - \bar{x})^2$	$v_t = n \cdot m - 1$		

[a] Here m = number of procedures; n = number of laboratories; $x_{i,j}$ = measurement of lab i and method j.

Table 4.14. Data Table of Dimension $n \times p$

	1	2	3	$\cdots$	p
1	x_{11}	...	...	...	x_{1p}
2	...	...	...	...	...
3	...	...	...	...	...
...	...	...	...	...	...
n	x_{n1}	...	...	...	x_{np}

are measured in function of time to verify whether the process still behaves as expected or whether it is drifting out of control.

A data table such as Table 4.14 is called a data matrix of dimension $n \cdot p$: n is the number of rows, representing the samples or objects, and p is the number of columns, representing the measurements or variables or parameters.

It is possible to do univariate statistics on each of the parameters separately. This will certainly yield some information. It is however clear that more information is hidden in the data table. For this it is important to investigate the simultaneous relationship between all the parameters in order to find the underlying structure. There is a broad range of multivariate statistical techniques available that can be used for this. A basic multivariate technique to investigate interdependencies among variables is principal components analysis (PCA). In the past few years this technique and its concepts have become very important in analytical chemistry. The reason for this is clearly the large amount of data that have become available as explained above and the availability of easy-to-use software packages. Because PCA is a basic technique that makes use of the major multivariate concepts it will be described in this chapter. Many techniques for modeling and multivariate calibration are based on PCA. We recommend several references for general reading on this subject (DI 84, MA 88b, MO 76) and for applications (CO 85, GR 89, KO 91, KV 88, MA 88a, MA 89, SK 92).

4.7.2. Principal Components Analysis as a Display Technique

The importance of visual inspection of data cannot be overemphasized when one is analyzing data. If contributes in an important way to better

Figure 4.23. A one-dimensional plot.

understanding of the obtained data. If only one measurement is available per sample, e.g., the concentration of one element, a plot can be drawn of that measurement, which can be seen as one variable (X_1); a one-dimensional plot as in Figure 4.23 is then obtained. The number of variables used to make the plot defines the dimensionality of the plot.

If in each sample a second element is analyzed, this concentration can be considered as a second variable (X_2); it is then again straightforward to draw two-dimensional plots of X_1 versus X_2 such as in Figure 4.24. In this figure it is clear that the concentration of the two elements are clearly related to each other. Moreover it can be seen that sample order is related to the concentration of the elements. The concentration of the elements increases with the sequence number.

Three-dimensional plots as in Figure 4.25 are still possible to draw and interpret. When one wants to visualize data such as those in Table 4.14, the matter becomes more complex.

To draw plots of more than three (e.g., k) variables requires k dimensions, and that is not possible. One solution can be to draw all or at least some of two- or three- dimensional plots of combinations of selected variables (e.g., X_1, X_2, and X_3). The drawback of this strategy is that a global overview of the data is never obtained and interpretation becomes quite difficult. A better approach is to use a data

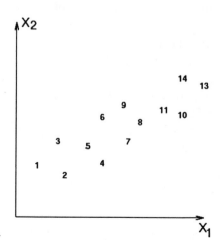

Figure 4.24. A two-dimensional plot.

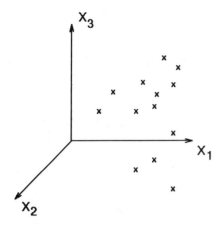

Figure 4.25. A three-dimensional plot.

reduction strategy that is able to reduce the dimensionality of the data to two or three.

PCA is essentially such a data reduction technique: it tries to minimize the dimensionality of the data matrix while still preserving as much information as possible. This is done by transforming the original variables. A set of linear combinations of the original variables is constructed in such a way that as much of the original variation of the data matrix as possible is preserved. The goal of PCA is to determine a few (usually two or three) factors (principal components) containing as much information as possible from the original data. The procedure is as follows:

The first component, denoted as $PC_{(1)}$, is the best linear combination of the original variables—best in the sense that this combination accounts for the largest amount of variation in the data among all possible combinations:

$$PC_{(1)} = w_{1(1)} X_1 + w_{2(1)} X_2 + \cdots + w_{p(1)} X_p \qquad (4.127)$$

The weights $w_{i(1)}$ define the best combination of the original variables, X_i.

The second principal component, $PC_{(2)}$, is that linear combination of the observed variables or measurements that accounts for the maximal amount of variation in the data that is not already accounted for by the first principal component. This means, in fact, that the second principal component is orthogonal to the first principal component. In general, the jth principal component is defined as that combination of the variables that accounts for the largest amount of variance not already explained by the previous components:

$$PC_{(j)} = w_{1(j)} X_1 + w_{2(j)} X_2 + \cdots + w_{p(j)} X_p \qquad (4.128)$$

In principle it is possible to extract as many principal components (PCs) as there are observed variables (X's). Since PCA is essentially a data reduction technique, it is our intention to select the first few (2 or 3) components that explain a large amount of the variation in the data.

To illustrate how PCA can be used for data reduction we come back to Figure 4.24. This is a two-dimensional plot, and data reduction is actually senseless. Imagine, however, that we anyway want to reduce the dimensionality to one. A linear combination of variables can be seen as a line like that drawn in Figure 4.26. If we consider the projections of the samples on that line, it is evident that this is not a good combination. Important information is lost. The relation for example between the sequence number and the concentration is lost. In Figure 4.27 a more logical combination is chosen. The variation in the original data is preserved as much as possible, and projections on this line still reveal the relationship between the sequence number and the concentration. This combination is exactly the first principal component, $PC_{(1)}$. Since there are two variables

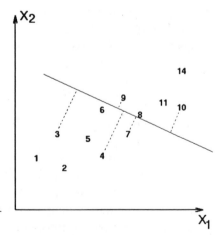

Figure 4.26. An arbitrarily line or combination of variables

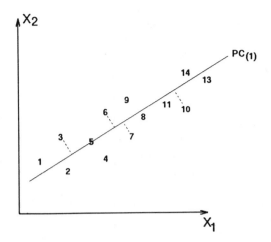

Figure 4.27. First principal component.

in this example, two principal components can be extracted. The direction of the second principal component is the line (or combination of variables) that is orthogonal to the first (see Figure 4.28).

It can be seen that the projection of the data on this component does not yield any relevant information. It most probably contains only the measurement error in the data. The two principal components can be seen as new coordinates in which the data can be

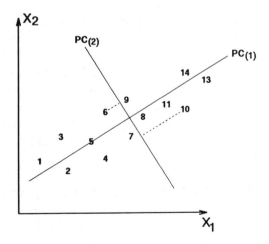

Figure 4.28. First and second principal components.

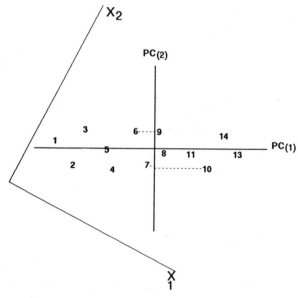

Figure 4.29. Projection of data in the coordinate axes of the principal components.

plotted. This is illustrated in Figure 4.29. In this case 100% of the variation is described since we have two variables and have extracted two principal components. In Figure 4.29 it can be seen that the major part of the variation resides along the axis of $PC_{(1)}$. This means that we have succeeded in reducing the dimensionality of these data from two to one. The first principal component, $PC_{(1)}$, can be seen as a new variable or feature that is a combination of the observed variables.

In general the PCA technique can be applied to reduce a data matrix of dimension p. Usually the first few principal components account for a large amount of the variation in the data so that visualization of the data by making plots of the first two or three components gives a good insight into and understanding of the structure of the data.

4.7.3.   Extraction of Principal Components

4.7.3.1.   Data Input for PCA

In PCA we describe the total variance of a data matrix of n objects (samples) and p variables (measurements) by constructing a new set of p

Table 4.15. Variance–Covariance Matrix

	1	2	3	$\cdots$	p
1	S_{11}	$\cdots$	$\cdots$	$\cdots$	S_{1p}
2	$\cdots$	S_{22}	$\cdots$	$\cdots$	$\cdots$
3	$\cdots$	$\cdots$	S_{33}	$\cdots$	$\cdots$
$\cdots$	$\cdots$	$\cdots$	$\cdots$	$\cdots$	$\cdots$
p	S_{1p}	$\cdots$	$\cdots$	$\cdots$	S_{pp}

linear orthogonal combinations of the observed variables. Essentially what is done is a reorganization of the available variance. In the original data matrix this variance is dispersed over all the variables. The total variance of the data matrix is given by the variance–covariance matrix **S**. The structure of **S** is given in Table 4.15.

The main diagonal elements are the variances of the individual variables, s_i^2. The off-diagonal elements are the covariances between the respective variables. The most common way to extract principal components starts from either the variance–covariance matrix or the correlation matrix (**R**). The structure of the correlation matrix is given in Table 4.16. When the original data are normalized to standardized scores (all variables have a mean 0 and a variance of 1), **S** = **R**. In computing correlation coefficients the differences in the means and variances between the variables are eliminated. The computations of (co-)variances removes only differences in the means of the variables.

Whether the data must be transformed to standard scores or not depends mainly on the data set itself. When there are large discrepancies between the variances of the variables, it is wise to use the standard scores. Otherwise the variables with the largest variances will dominate the analysis. By transforming them to standard scores (or, equivalently,

Table 4.16. Correlation Matrix

	1	2	3	$\cdots$	p
1	1.0	$\cdots$	$\cdots$	$\cdots$	r_{1p}
2	$\cdots$	1.0	$\cdots$	$\cdots$	$\cdots$
3	$\cdots$	$\cdots$	1.0	$\cdots$	$\cdots$
$\cdots$	$\cdots$	$\cdots$	$\cdots$	$\cdots$	$\cdots$
p	r_{1p}	$\cdots$	$\cdots$	$\cdots$	1.0

using **R**) they all get equal importance. Unfortunately data scaling has an influence on the results and interpretation of the results of PCA and all multivariate techniques in general. The fact that there is no easy way to translate the results between the scaling methods makes the problem even more seriously. The use of standard scores makes the interpretation usually somewhat easier. It is common practice to use standardized scores unless the variance of the variables does not differ too much.

The total variance in the data set is given by the trace of **S** (or **R**). This is the sum of the diagonal elements or, in the case of **S**, the sum of the variances of the individual variances. Each variable contributes to the total variance through its own variance, s_i^2.

The trace of **R** equals the total variance of the standardized data. Since the variables are transformed to have all unit variance, the total variance equals the number of variables ($=$ trace of **R**).

4.7.3.2. *Extraction of Principal Components from the Variance–Covariance Matrix*

In what follows we assume that all data are mean corrected.

As explained before, the aim of PCA is to construct a set of new orthogonal variables. The total variance is concentrated in the first few variables. The variance–covariance matrix of the principal components is a diagonal matrix (see Table 4.17).

The off-diagonal elements are zero since the principal components are orthogonal. The diagonal elements are the variances of the respective principal components. Ideally the first two or three are much larger than the others, so that only those can be preserved. More about the number of factors to be retained can be found in Section 4.7.4.1.

Table 4.17. Variance–Convariance Matrix of Principal Components

	1	2	3	$\cdots$	p
1	S_{11}				0.0
2		S_{22}			
3			S_{33}		
⋮				$\cdots$	
p	0.0				S_{pp}

Calculating the first principal components means calculating weights $w_{(1)}$ of eqs. 4.127, 4.128 and 4.129:

$$PC_{(1)} = x \cdot w_{(1)} \qquad (4.129)$$

where $w_{(1)}$ is the $(p \times 1)$ vector of the weights that has to be chosen to maximize the variance of the constructed principal component. This variance is expressed as

$$\frac{1}{n-1}(Xw)' \, Xw = \frac{1}{n-1} w'X'Xw = w'Sw \qquad (4.130)$$

subject to $w'w = 1$. This constraint—that the sum of squared coefficients must equal 1—is used to prevent the coefficients from becoming arbitrarily high. Indeed, multiplication of the coefficients of a certain weight vector, w, does not change the direction in the space, but it does increase the variance as expressed in eq. 4.130. Since the direction is the basic characteristic we are looking for, the coefficients must be normalized. Usually they are normalized to 1.

It can be shown that the eigenvectors (or characteristic vectors) of S are the weight vectors w that define the principal components and that the associated eigenvalue γ_1 is the variance of the principal component.

The first eigenvector of S is defined as

$$Sw = \lambda_1 w_{(1)}$$
$$(S - \lambda_1 I)w_1 = 0 \qquad (4.131)$$

where w_1 is the first eigenvector and λ_1 is the associated eigenvalue. If the solution of this equation is to be other than the null vector, λ_1 must satisfy

$$|S - \lambda_1 I| = 0 \qquad (4.132)$$

where λ_1 is thus the largest eigenvalue of S (Table 4.18).

Principal components are orthogonal to each other. The variance–covariance matrix of the principal components is therefore a diagonal matrix, Λ (see Table 4.17). The diagonal elements are the respective variances of the principal components, the eigenvalues of S

Table 4.18. Principal Components

The first principal component of the observable variables (measurements) $X_{1,...,p}$ is the linear combination:

$$PC_{(1)} = w_{(1)1}X_1 + w_{(1)2}X_2 + \cdots + w_{(1)p}X_p$$

whose coefficients are the elements of the first eigenvector of the variance–covariance matrix **S**, associated with the largest eigenvalue, λ_1. The coefficients $w_{(1)}$ are unique up to a scale factor. If they are scaled to length 1 ($w'w = 1$), then the associated eigenvalue λ_1 is the variance of this principal component.

Or, in general, the jth principal component of the observable variables (measurements) $X_{1,...,p}$ is the linear combination:

$$PC_{(j)} = w_{(j)1}X_1 + w_{(j)2}X_2 + \cdots + w_{(j)p}X_p$$

whose coefficients are the elements of the jth eigenvector of the variance covariance matrix **S**, associated with the jth largest eigenvector, λ_j. If they are scaled to length 1 ($w'w = 1$), then the associated eigenvalue λ_j is the variance of this principal component. The principal components are necessarily orthogonal to each other.

---

in decreasing sequence. The total variance in the data set is thus

$$\lambda_1 + \lambda_2 + \lambda_3 + \cdots + \lambda_p = \text{trace } \Lambda = \text{trace } \mathbf{S} = \sum_{i=1}^{p} s_i^2 \quad (4.133)$$

The relative contribution of each principal component to the total variance is measured by

$$\frac{\lambda_i}{\text{trace}(\Lambda)} = \frac{\lambda_i}{\text{trace}(\mathbf{S})} \quad (4.134)$$

Since the principal components are new variables, derived from the original measurements on the samples, a value for each sample for these components can be calculated. These values are called the component scores, $\mathbf{F}_{i(j)}$,

$$\mathbf{F}_{i(j)} = \mathbf{w}_{1,(j)}\mathbf{X}_{i1} + \mathbf{w}_{2,(j)}\mathbf{x}_{i2} + \cdots + \mathbf{w}_{p,(j)}\mathbf{X}_{ip} \quad (4.135)$$

which is the score for the ith object on the jth principal component (X_{ip} is the value of the ith object for the pth variable).

The coefficients $w_{p(1)}$ indicate the importance of the pth variable on the first PC. Both the sign and the magnitude are relevant. This can be more precisely quantified by calculating the correlations between the original variables and the principal components. These are often called *component loadings*. It can be shown that they can be calculated as follows (MO 76):

$$\mathbf{l}_{i,(j)} = \mathbf{w}_{i,(j)} \frac{\sqrt{\lambda_{(j)}}}{\sqrt{\mathbf{S}_{(i,i)}}} \qquad (4.136)$$

In some studies component loadings are defined as

$$\mathbf{l}_{i,(j)} = \mathbf{w}_{i,(j)} \sqrt{\lambda_j} \qquad (4.137)$$

When one is using standardized variables these two definitions are identical since all variables then have unit variance.

4.7.3.3. Extracting Principal Components from the Correlation Matrix

When the measurements are standardized to zero mean and unit variance then $\mathbf{S} = \mathbf{R}$. Extraction of principal components from $\mathbf{R}$ is analogous to the extraction of the components from $\mathbf{S}$. It suffices to replace $\mathbf{S}$ by $\mathbf{R}$. The total variance now equals p, the number of variables:

$$\text{trace}(\mathbf{R}) = p = \sum_{i=1}^{p} s_i^2 \qquad (4.138)$$

The total sum of the eigenvalues also equals p:

$$\lambda_1 + \lambda_2 + \lambda_3 + \cdots + \lambda_p = \text{trace}\,\mathbf{\Lambda} = \text{trace}(R) = \sum_{i=1}^{p} s_i^2 = p \qquad (4.139)$$

The contribution of the jth component to the total variance is now

$$\frac{\lambda_j}{p} \qquad (4.140)$$

The component score of the ith object on the jth principal component is

$$\mathbf{F}_{i,(j)} = \mathbf{w}_{1,(j)}\,\mathbf{Z}_{i1} + \mathbf{w}_{2,(j)}\,\mathbf{Z}_{i2} + \cdots + \mathbf{w}_{p,(j)}\,\mathbf{Z}_{ip} \qquad (4.141)$$

The $\mathbf{Z}$ values are the standardized measurements.

The correlation coefficient of each principal component with the original measurement is

$$\mathbf{I}_{i,(j)} = \mathbf{w}_{i,(j)}\,\sqrt{\lambda_j} \qquad (4.142)$$

4.7.3.4. Q-Type Principal Components

We have already considered the covariances or correlations between the variables and extracted components from the variance–covariance matrix or correlation matrix. This approach is historically called R-type analysis. Although this is the most common type of analysis, it is of course also possible to extract components from the covariances or correlation matrix between the objects. This is called Q-type factor analysis. Although this type of analysis results in the same eigenvalues, the associated eigenvectors are different.

It is possible to express the original matrix, $\mathbf{X}$, in terms of the eigenvectors of its R-type analysis, the eigenvectors of the Q-type analysis, and a diagonal matrix of so-called singular values. This is called the *singular value decomposition* of a data matrix:

$$\mathbf{X} = \mathbf{U} \cdot \mathbf{P} \cdot \mathbf{V}' \qquad (4.143)$$

where $\mathbf{X}$ is the $n \times p$ matrix of the measurements, usually $n \leqslant p$

$\mathbf{V}$ is the $p \times (p)$ matrix of eigenvectors of the covariances between the measurements (R-type analysis); the columns of $\mathbf{V}$ are the eigenvectors

$\mathbf{U}$ is the $n \times (p)$ matrix of eigenvectors of the correlation matrix between the objects (Q-type analysis components); the columns of $\mathbf{U}$ are the eigenvectors.

$\mathbf{P}$ is the $(p) \times (p)$ diagonal matrix of singular values; the diagonal elements or the singular values are the square roots matrix of the eigenvalues that are common for Q-type and R-type analysis

4.7.4. Interpretation of Principal Components

Principal components analysis as a data reduction technique is often used as a first step in analysis of multivariate data. The ability to visualize the data in two or three dimensions is therefore of the utmost importance. There are basically two sorts of useful plots that can be obtained from a PCA. The first reveals the interrelation between the samples. It is the plot of the component scores of the samples of objects in the new coordinate axes of the first two principal components. This was discussed earlier and illustrated in Figure 4.29.

It is however also possible to reveal the interrelationships between the measurements. This can be accomplished by inspecting or plotting the loadings of the variables of the first two principal components. This is illustrated in Figure 4.30. Variables that have high loadings on a principal component contribute greatly to the variance accounted for by this component, whereas variables with small loadings do not. This enables one to make an interpretation of the principal components and obtain a better insight into the underlying structure of the data. Variables that have high loadings on the same principal components are related. This reveals relationships between the measurements.

Related multivariate techniques like correspondence factor analysis allow the combination of both plots in one, called *biplots*. Relationships between objects and variables can be more easily derived from these techniques.

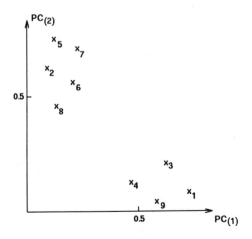

Figure 4.30. Plot of the loadings of the variables.

4.7.4.1.   Number of Components to Retain

In the previous sections we implicitly assumed that two or three principal components have to be considered. When the aim is visualization, this is certainly logical. Sometimes it may be interesting to know how many principal components are important. This can be the case when one is going to do further analysis, based on the, say, q first principal components. It is common practice to take the number of components that account for a "large" part of the variance, e.g., $> 75\%$ of the total variance. Some statistical tests of their significance are available. One should, however, be aware that statistical significance results in the retaining of a rather large number of components. This may neither be practical nor have chemical significance. Therefore several heuristic rules are proposed. One that is often used is Kaiser's rule, which states that all components with an eigenvalue larger then 1 should be retained. The rationale behind this is that a component should account for more variance then the original variables, which were scaled to unit variance. Another approach is the scree test. Here the eigenvalues are plotted against their sequence number (see Figure 4.31).

The components with small variance fall on a straight line, whereas the first components clearly lie above this line. Components before the straight line are then retained. This is, however, not always possible to determine unambiguously. There may be no clear elbow in the curve, or serve nicks may occur (Figure 4.32). Then the test remains inconclusive. Several variations have been proposed, but in general there is as yet no

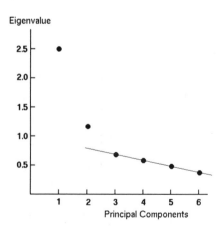

Figure 4.31. Scree plot of the eigenvalues of the principal components.

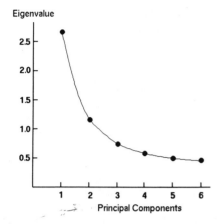

Eigenvalue

Figure 4.32. Problematic scree plot.

clear criterion for this problem and decisions will usually be based on common sense.

Example. A lake has been sampled and a number of heavy metals, PCBs and PAHs have been analyzed (BO 73, BO 79). The resulting data matrix is shown in Table 4.19. The rows (the objects) represent the sampling points. The columns (the variables) represent the different analyses.

The PCA analysis involves the steps:

1. Scaling the data; in this example standardization is carried out
2. Calculating the correlation matrix, **R**
3. Extracting the eigenvectors and eigenvalues of **R**

The results are shown in the following tables:

Table 4.20: The standardized data
Table 4.21: The correlation matrix, **R**
Table 4.22: Matrix of eigenvectors of **R**
Table 4.23: Matrix of eigenvalues

The first principal component is then defined by the first eigenvector (= column 1 of Table 4.22).

$$PC_{(1)} = 0.42X_1 + 0.40X_2 + \cdots + 0.25X_8 \qquad (4.144)$$

Since there are 8 variables, 8 PCs can be calculated.

Table 4.19. Concentration of Different Pollutants in the Lake Sampling Example

Dry Soil (mg/kg)				Dry Soil $(\mu g/kg)^a$			
Cu	Cr	Cd	Pb	PCB_{28}	PCB_{52}	Bghipe	Inp
149.00	208.00	3.70	159.00	127.00	114.00	633.00	506.00
127.00	160.00	3.30	138.00	84.00	72.00	602.00	602.00
104.00	145.00	2.70	89.00	67.00	60.00	746.00	522.00
168.00	255.00	5.90	193.00	103.00	103.00	619.00	515.00
239.00	397.00	7.60	400.00	116.00	129.00	838.00	1500.00
259.00	489.00	9.30	392.00	178.00	147.00	1304.00	1007.00
222.00	431.00	6.20	291.00	246.00	148.00	804.00	640.00
163.00	352.00	5.60	270.00	61.00	76.00	1069.00	763.00
177.00	340.00	5.90	221.00	183.00	183.00	427.00	366.00
154.00	331.00	3.60	178.00	111.00	99.00	432.00	309.00
216.00	456.00	9.00	313.00	133.00	133.00	619.00	531.00
143.00	286.00	3.40	182.00	73.00	55.00	366.00	305.00
226.00	451.00	7.50	294.00	174.00	174.00	435.00	304.00
242.00	485.00	8.70	319.00	182.00	136.00	545.00	409.00
206.00	430.00	6.90	292.00	149.00	149.00	693.00	891.00
257.00	553.00	9.80	367.00	156.00	156.00	1016.00	1016.00
256.00	572.00	8.90	349.00	198.00	159.00	1032.00	952.00
279.00	626.00	10.60	389.00	181.00	181.00	927.00	847.00
202.00	400.00	7.70	290.00	118.00	118.00	630.00	472.00
212.00	392.00	6.80	294.00	181.00	120.00	964.00	723.00
185.00	380.00	7.70	280.00	115.00	115.00	992.00	916.00
224.00	470.00	6.80	291.00	142.00	94.00	943.00	943.00
166.00	351.00	5.60	256.00	88.00	88.00	789.00	702.00

^{a} PCB_{28} and PCB_{52} = polychlorinated biphenyls; Bghipe = Benzo[ghi]perylene; Inp = Indenoperylene.

The matrix of the values of the objects on $PC_{(1)}$, the scores, are shown in Table 4.24. The rows represent the scores (values) of the sampling points on the different PCs.

The eigenvalues are the variances of the associated eigenvectors. The variance of the first PC = 5.47 = the first eigenvalue. This can be verified by calculating the variance of the scores on $PC_{(1)}$:

$$\frac{1}{22} \sum_{i=1}^{23} F_{ij}^2 \tag{4.145}$$

The total variance in the data set is

$$\text{trace}(\mathbf{R}) = 8 = \text{trace } \mathbf{\Lambda} = \text{the number of variables}$$

Table 4.20. Standardized Values for the Data of Table 4.19 (× 100)

X_1	X_2	X_3	X_4	X_5	X_6	X_7	X_8
− 106.56	− 146.86	− 132.30	− 133.71	− 22.26	− 21.75	− 50.79	− 60.39
− 153.48	− 185.69	− 150.17	− 158.64	− 112.09	− 134.11	− 63.42	− 27.89
− 202.54	− 197.82	− 176.98	− 216.83	− 147.61	− 166.22	− 4.73	− 54.98
− 66.03	− 108.84	− 34.00	− 93.34	− 72.40	− 51.18	− 56.49	− 57.35
85.41	6.01	41.96	152.45	− 45.24	18.38	32.76	276.12
128.07	80.43	117.92	142.95	84.30	66.53	222.69	109.22
49.15	33.52	− 20.59	23.02	226.37	69.21	18.91	− 15.03
− 76.69	− 30.39	− 47.40	− 1.91	− 160.15	− 123.41	126.91	26.61
− 46.83	− 40.09	− 34.00	− 60.09	94.74	162.84	− 134.74	− 107.79
− 95.89	− 47.37	− 136.77	− 111.15	− 55.68	− 61.88	− 132.70	− 127.09
36.35	53.74	104.52	49.15	− 9.72	29.08	− 56.49	− 51.93
− 119.35	− 83.77	− 145.70	− 106.40	− 135.07	− 179.59	− 159.60	− 128.44
57.68	49.69	37.49	26.59	75.94	138.77	− 131.48	− 128.78
91.81	77.19	91.11	56.27	92.65	37.10	− 86.65	− 93.23
15.02	32.71	10.68	24.21	23.71	71.88	− 26.33	69.95
123.80	132.20	140.26	113.27	38.33	90.61	105.31	112.26
121.67	147.56	100.05	91.89	126.08	98.64	111.83	90.60
170.73	191.24	176.01	139.39	90.56	157.49	69.04	55.05
6.49	8.44	46.43	21.84	− 41.04	− 11.05	− 52.01	− 71.90
27.82	1.97	6.22	26.59	90.56	− 5.70	84.12	13.07
− 29.77	− 7.74	46.43	9.96	− 47.33	− 19.08	95.53	78.41
53.42	65.06	6.22	23.02	9.08	− 75.26	75.56	87.55
− 70.29	− 31.19	− 47.40	− 18.53	− 103.74	− 91.31	12.79	5.96

Table 4.21. Correlation Between the Pollutant Concentrations

	X_1	X_2	X_3	X_4	X_5	X_6	X_7	X_8
X_1	1.00	.95	.93	.95	.73	.77	.47	.51
X_2	.95	1.00	.91	.90	.67	.72	.44	.41
X_3	.93	.91	1.00	.92	.59	.73	.49	.47
X_4	.95	.90	.92	1.00	.56	.66	.56	.65
X_5	.73	.67	.59	.56	1.00	.83	.19	.11
X_6	.77	.72	.73	.66	.83	1.00	.12	.18
X_7	.47	.44	.49	.56	.19	.12	1.00	.76
X_8	.51	.41	.47	.65	.11	.18	.76	1.00

The proportion of the total variance that is explained by the different PCs is given in Table 4.25.

As can be seen, the first two components explain 87.5% of the variance. A scree plot (Figure 4.33) of the eigenvalues confirms that two or three components are significant.

Table 4.22. Eigenvectors of the Data of Table 4.20 ($\times 100$)[a]

	$PC_{(1)}$	$PC_{(2)}$	$PC_{(3)}$	$PC_{(4)}$	$PC_{(5)}$	$PC_{(6)}$	$PC_{(7)}$	$PC_{(8)}$
X_1	41.92	−6.30	13.10	−1.54	26.71	−19.92	10.58	−82.50
X_2	40.10	−8.52	30.39	−28.42	23.84	73.97	14.81	18.14
X_3	40.32	−2.00	35.33	−12.06	−41.30	−45.00	48.32	30.18
X_4	40.88	10.41	27.51	11.25	17.98	−25.09	−75.44	26.34
X_5	31.25	−39.87	−65.27	−13.06	39.51	−21.91	12.33	28.46
X_6	33.79	−39.53	−23.97	39.84	−61.15	29.14	−18.97	−13.63
X_7	24.39	57.00	−41.72	−54.42	−32.18	5.94	−15.89	−11.45
X_8	25.00	58.09	−18.01	64.79	18.82	10.60	29.64	11.54

[a] The columns represent the eight principal components; the rows represent the eight variables X_1 to X_8.

Table 4.23. Variance–Covariance Matrix of the Principal Components

PC	1	2	3	4	5	6	7	8
1	5.47	0	0	0	0	0	0	0
2	0	1.53	0	0	0	0	0	0
3	0	0	0.44	0	0	0	0	0
4	0	0	0	0.28	0	0	0	0
5	0	0	0	0	0.15	0	0	0
6	0	0	0	0	0	0.07	0	0
7	0	0	0	0	0	0	0.004	0
8	0	0	0	0	0	0	0	0.001

The loadings, as defined in eq. 4.137 for the variables on the principal components, are shown in Table 4.26. A plot of the loadings shows that the variables cluster into three main groups, the heavy metals, the PAHs, and the PCBs (Figure 4.34). All variables have positive loadings on $PC_{(1)}$. This means that $PC_{(1)}$ can be seen as an indication of overall pollution. This PC also separates the inorganic heavy metals contaminants from the organic ones. On $PC_{(2)}$ the distinction between the organic pollutants can be seen.

4.8.   PATTERN RECOGNITION

Pattern recognition may be described as a tool designed to aid the scientist in making an educated guess (DE 78) in the systematic analysis

Table 4.24. Scores of the Objects (Rows) on the Eight PCs (Columns) (× 100)

$PC_{(1)}$	$PC_{(2)}$	$PC_{(3)}$	$PC_{(4)}$	$PC_{(5)}$	$PC_{(6)}$	$PC_{(7)}$	$PC_{(8)}$
−253.35	−38.61	−90.31	27.05	−23.40	−5.19	−4.54	−18.39
−366.98	57.33	−36.44	33.05	1.10	−20.63	16.79	−3.74
−441.41	100.51	−60.75	−23.70	−32.67	5.87	25.09	−4.46
−191.22	−12.02	14.00	8.20	−36.26	−37.12	23.63	−13.89
186.55	198.98	31.50	183.36	46.65	−23.23	−17.22	−3.84
322.36	128.06	−61.34	−57.91	−28.09	−29.24	−30.63	−15.81
130.12	−118.70	−153.84	−27.22	71.97	−11.41	−9.83	16.48
−118.37	208.60	39.82	−64.77	−32.13	24.09	−42.70	2.73
−49.17	−240.77	−72.11	65.47	−57.91	17.33	−11.74	13.05
−262.22	−101.56	23.58	−8.57	34.21	46.32	−11.12	−6.90
79.05	−73.83	103.83	−12.95	−25.02	−25.12	12.11	17.99
−359.79	−34.24	99.09	−19.02	72.82	8.25	−10.48	0.77
76.45	−240.73	38.43	16.95	−20.23	4.07	−22.71	−17.36
126.24	−163.48	66.72	−38.57	39.67	−40.83	13.25	1.05
76.38	−13.67	−11.98	77.10	−1.20	31.92	7.54	11.17
304.11	64.06	26.21	2.87	−25.99	17.74	19.10	−6.55
310.73	14.44	−47.50	−24.87	13.46	34.19	19.68	0.87
388.32	−43.09	45.43	−13.51	−28.80	29.25	12.94	−13.58
−13.46	−50.46	89.92	−22.97	−17.77	−26.36	−8.12	3.42
75.98	22.41	−81.44	−50.16	24.77	−28.68	−11.06	4.71
28.88	129.05	−5.62	−4.43	−50.19	−4.40	16.47	27.03
78.12	113.42	0.12	−33.04	73.12	18.75	30.27	−10.04
−127.32	94.31	42.65	−12.36	1.88	14.42	−16.72	15.29

Table 4.25. Proportion of Variance Accounted for by the Different Components

PC	Variance	Proportion of Variance (%)	Cumulative Proportion
1	5.47	68.4	68.4
2	1.53	19.1	87.5
3	0.44	5.6	93.1
4	0.28	3.5	96.6
5	0.15	1.9	98.5
6	0.065	0.8	99.3
7	0.004	0.5	99.8
8	0.0015	0.2	100.0

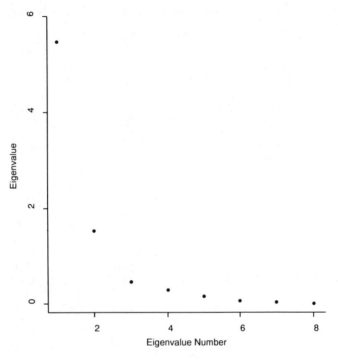

Figure 4.33. Scree plot for the lake pollution example.

Table 4.26. Loadings or Correlation Coefficients Between the Variables (Rows) and PCs (Columns) (× 100)

	PC$_{(1)}$	PC$_{(2)}$	PC$_{(3)}$	PC$_{(4)}$	PC$_{(5)}$	PC$_{(6)}$	PC$_{(7)}$	PC$_{(8)}$
X_1	98.05	−7.78	8.72	−0.82	10.66	−5.07	2.08	−10.26
X_2	93.78	−10.53	20.22	−15.10	9.51	18.83	2.91	2.26
X_3	94.30	−2.47	23.51	−6.41	−16.48	−11.45	9.49	3.75
X_4	95.61	12.86	18.30	5.98	7.17	−6.39	−14.81	3.28
X_5	73.08	−49.27	−43.43	−6.94	15.77	−5.58	2.42	3.54
X_6	79.02	−48.86	−15.95	21.17	−24.40	7.42	−3.73	−1.70
X_7	57.03	70.44	−27.76	−28.91	−12.84	1.51	−3.12	−1.42
X_8	58.46	71.79	−11.98	34.42	7.51	2.70	5.82	1.44

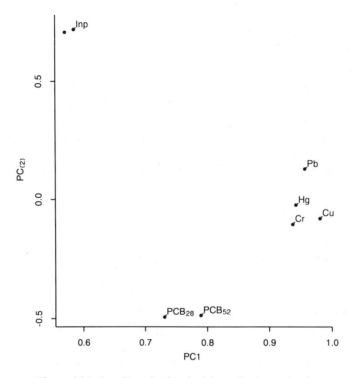

Figure 4.34. Loading plot for the lake pollution example.

of multidimensional data when direct or statistical analysis is not feasible and calculation based on a physical model is impossible. This is a rather general statement; more specifically, pattern recognition finds and/or predicts a property of the objects that is not measurable but is known to be related to the measurements via some unknown relationships, given a set of objects and a list of measurements made on those objects (KO 84).

In this description a property is the information (one hopes) implicit in the data one obtains, for example, the origin of some material (using elemental composition measurements), the quality of some material (using chemical and physical measurements), the structure of some compound (using spectral measurements), or the chemical and/or biological activity of some compound (using molecular structure and composition information). In this context properties can be considered to lie along a continuous scale ranging from *discrete category* to *continuous property*.

Discrete category data have as their property an *identifier* that may be arbitrarily assigned (material problems are often of this nature). The value of such a property is not intended to be a function of the measurements; it is assigned to permit identification of the discrete types or groups—categories—of objects. Continuous property data have a value that is assumed to be a function of the measurements. Real applications often fall between these extremes into what are termed continuous categories. Reactivity data taken as continuous values are often no more accurate than could be described by the words *low*, *middle*, and *high*. This range of possible property types would be of little importance except that the two extremes of discrete category and continuous property require different pattern recognition methods.

Generally, the use of discrete categories or continuous categories allows more, and more powerful, pattern recognition methods to be applied. This is intrinsic to the simpler "is it or is it not?" question that the categories pose, in comparison to the question of "how much" posed with continuous property. However, the arbitrary nature of the category identifier in a truly discrete category situation makes the analysis of that problem with continuous property methods meaningless. A reassignment of the category identifiers gives very different results. (An exception is the situation in which only two categories are present: switching the assignment merely interchanges the signs of the results.)

The approach based on pattern recognition algorithms often gives good results, but it cannot be considered universal, since in each particular case the computer has to recognize some new classes of compounds. As a rule, pattern recognition algorithms are used to solve only individual problems, without any real claim to versatility.

Computer learning aimed at the recognition of a new class of compounds cannot be carried out successfully unless the training data are adequate and representative. The narrow specialization of the identification systems incorporating pattern recognition algorithms is a major disadvantage of the aforementioned approaches.

4.8.1. Data Collection and Presentation

The measured data on objects may result from various one-dimensional analytical procedures such as melting point, boiling point, or refraction measurement. Here the set of measurements is termed *multi-source data*. Measured data exclusively resulting from one two-dimensional method such as spectral analysis are named *single-source data*. Here the two dimensions are spectral position and intensity. Both types of data may be used in pattern recognition procedures.

The collecting of multisource data is done by making tables, or data records in an input file of a computer. Single-source data can automatically be collected by means of on-line dedicated computers belonging to the measuring apparatus. When the signal-to-noise ratio is a problem, single-source data may be sampled many times and coherently added. Modern spectroscopic techniques have built-in facilities for such a procedure and the signal-to-noise improvement is visually controllable.

In order to apply pattern recognition, all measured data on one object are conveniently represented as a vector $\mathbf{x}$ of dimension n in a hypergeometric space R^n. Here n equals the dimensionality and the number of measured properties per object.

A spectrum of a compound can be represented as a number of frequencies n, each with an intensity I_n. The collection of n intensities provides one compound vector $\mathbf{x}_n$. Comparison of two different spectra can be considered as measuring the distance between the endpoints of two vectors $\mathbf{x}_i$ and $\mathbf{x}_j$ of the same dimensionality in the representative space R^n.

Pattern recognition is applied in the process of comparing a large collection of representation points in order to detect clusters of points. Such a cluster is attributed to objects with like properties.

4.8.2. Preprocessing Techniques

Sometimes the measured data forming n-dimensional patterns in the hypergeometric space are subjected to transformation procedures before the actual pattern recognition techniques are applied. These transformations serve the following purpose:

- The separation of various clusters of like vectors is improved in order to simplify the classification.
- The dimensionality of the pattern space R^n is reduced in order to economize on the computing time and to visualize the result.

One method of preprocessing that may be an advantage is auto-scaling. This technique distributes the measured data in such a manner along the axis of the coordinate system that the average value of the coordinates along each of the axes equals 0 and the variance equals 1. This procedure aims for an equal resolution along each of the axis. The values of the coordinates in the pattern space R^n thus change by autoscaling. The jth coordinate value of pattern vector **a** becomes

$$x'_{aj} = \frac{x_{aj} - \bar{x}}{\sigma} \qquad (4.146)$$

where x'_{aj} represents the autoscaled jth coordinate of pattern x_{aj}, the original value; $\bar{x}_j = (1/N)\sum x_{aj}$, the average value of j for all N patterns; and $\sigma_j^2 = \sum (x_{aj} - \bar{x}_j)^2/(N-1)$ the variance of the j-coordinate of all patterns.

4.8.3.   Recognition Without Supervision

Recognition here is the discovery of clusters of patterns in the n-dimensional hypergeometric space R^n. In situations where n equals 2 or 3, the technique for recognition that is applied most successfully is visual recognition because human visual perception is a very powerful method. When the dimensionality of the hyperspace exceeds 3, two ways are open:

- Dimension reduction
- Mathematical recognition methods

Dimension reduction may proceed in two different ways:

1. Linear combinations of features on the different axes in the n-dimensional space may be constructed in such a way that n features produce n orthogonal linear combinations. These orthogonal combinations are ordered according to decreasing variance. Because the variance of a particular feature—or feature combination—contains the information content of these features for the proper description of the patterns, the combination with the highest variance is the most significant for the pattern description. In practical situations it sometimes occurs that the two or three topmost feature combinations—"eigenvectors"—encompass a considerable percentage of all available information (sum of "eigenvalues").

In such situations one may depict these two or three eigenvectors along three orthogonal axes in order to survey the entire problem without causing too much uncertainty by omitting the other eigenvectors. Hereafter one may rely on visual inspection.

2. In the nonlinear mapping procedure, all points in the R^n space are projected into the R^2 space under the condition that the mutual distance between each pair of points remains the same. When the directions along the two axes in the R^2 space are called x and y, this condition for two points a and b implies

$$D_{ab} = D_{ab}^* = [(x_a - x_b)^2 + (y_a - y_b)^2]^{1/2} \qquad (4.147)$$

where D_{ab} equals the distance between a and b in R^n, and D_{ab}^* equals the distance between a and b in R^2. For many points these distances cannot all be transformed without making errors; therefore an error is defined according to

$$E = \frac{\sum\limits_{a>b} (D_{ab} - D_{ab}^*)^2}{D_{ab}} \qquad (4.148)$$

when the summation sums over all pairs of points and

$$D_{ab} = \left[\sum\limits_{j=1}^{n} (x_{a,j} - x_{b,j})^2 \right]^{1/2} \qquad (4.149)$$

This error E represents a nonlinear function of $2N$ unknowns (the x and y values of N points in the R^2 space). Minimalization of E with respect to all unknown produces the position of all points in R^2.

Mathematical recognition methods start from the position of all points in R^n. These positions are transformed into a distance matrix that lists all distances between all points or into an equality matrix G that related to the distance matrix according to

$$G_{a,b} = 1 - \frac{D_{a,b}}{D_{max}} \qquad (4.150)$$

where D_{max} equals the distance between the two points that are farthest apart. It is apparent that these two points produce an equality 0 whereas two points at the same location give $G = 1$.

A particular method of cluster search, called hierarchical or Q-mode clustering, is based on the equality matrix. Here the pair of points with highest equality are sought and substituted for their center of gravity. This center of gravity gets an associated weight constant, denoting the number of points involved in the substitution. This process is repeated for the remaining points until one of the following occurs:

- A number of clusters—or gravity centers—is obtained equal to a fixed number set in advance.
- The highest remaining equality value is lower than a limiting value set in advance.
- The decrease in equality on repetition of the clustering procedure surpasses a limiting value.

It should be mentioned here that instead of Euclidean distance matrices—with an exponent value $\mu = 2$ in $[\sum(x_{aj} - x_{bj})]^{1/\mu}$—sometimes other exponents and used (e.g., $\mu = 1$). The general name for such distances is Minkovski distances.

The position of the patterns in R^2 can be visualized by various methods. A very useful one is that using the line printer of a modern computer. Here the patterns, denoted by their number or name, are printed in a position corresponding with the x and y values in R^2. Instead of a pattern number, one may also use the class number in situations when supervised classification can be applied. A disadvantage can be that two or more patterns may be located at the same position.

4.8.4. Training and Classification

Up to now the recognition methods applied refer to situations in which it is not known whether or how many clusters exist. Sometimes the situation is not so bad, and the number of clusters or classes of compounds is known. Here pattern recognition methods with supervision may be applied that use the information on the number of classes given. Such procedures are called training and classification. In this context training pertains to the methods applied to classify the various clusters and it aims for the development of a strategy that can be successfully used when new unknown samples should be classified. Thus training is the development of decision operators using well-classified patterns aiming for correct classification of unknown patterns. Methods with supervision start with attributing weights to the various features, taking into account their value for class separation. This feature weighing is part of the preprocessing technique.

4.8.5. Feature Weighing

In a particular pattern recognition method, not all measured data—features—of an object are equally important for the description of the property associated with the patterns. In order to evaluate the individual importance of each feature, evaluation rules (weighing functions) should be applied to select the most important ones. Three different weighing procedures may be mentioned here.

Variance weighing is a procedure that measures the utility feature j has for the separation of category m from category n. The algorithm for the value of this utility $W(v)_{j,m,n}$ is

$$W(v)_{j,m,n} = \frac{[\overline{(x_m)_j^2} + \overline{(x_n)_j^2} - 2\overline{(x_m)_j(x_n)_j}]}{(m2)_{m,j} + (m2)_{n,j}} \qquad (4.151)$$

where $(m2)_{m,j}$ equals the second moment of feature j in class m $[= (N_{jn} - 1)\sigma_{m,j}^2/N_{jn}]$. The variance weight of feature j for all linear class separations $W(v)_j$ equals, for M classes,

$$W(v)_j = \left[\prod_{m=1}^{M-1} \prod_{n=m+1}^{M} W(v)_{j,m,n} \right]^{2/M(M-1)} \qquad (4.152)$$

An increase in correlation between the feature j of category m and n here results in a decrease in the utility of this feature for the separation.

Another weighing is that according to Fisher. Here the weighing factor equals

$$W_{j,m,n}(F) = \frac{[\overline{(x_m)_j} - \overline{(x_n)_j}]^2}{N_m \sigma_{m,j}^2 + N_n \sigma_{n,j}^2} \qquad (4.153)$$

The population of a class is taken into account in this equation. The

Fisher weight of feature j for all linear class separations $W(F)_j$ equals

$$W(F)_j = \frac{2\left[\displaystyle\sum_{m=1}^{M-1}\sum_{n=m+1}^{M} W(F)_{j,m,n}\right]}{M(M-1)} \tag{4.154}$$

A weighing factor that takes into account the value of feature j for separating property p_k from other properties is

$$W_j(p) = \frac{\left[\displaystyle\sum_{k=1}^{N} (x_{j,k} - \bar{x})(p_k - \bar{p})\right]^2}{\sigma_j^2(N-1)^2\sigma_p^2} \tag{4.155}$$

or, in another notation,

$$W_j(p) = \frac{\left[\displaystyle\sum_{k=1}^{N} (x_{j,k} - \bar{x})(p_k - \bar{p})\right]^2}{\displaystyle\sum_{k=1}^{N} (x_{j,k} - \bar{x}_j)^2 \sum_{k=1}^{N} (p_k - \bar{p})^2} \tag{4.156}$$

where N denotes the number of patterns in the entire training set; $\bar{x}_j$ equals the average value of feature j; and $\bar{p}$ is the average value of a property. In a shorter notation, with $x_{j,k} - \bar{x}_j = x_{j,k}^1 (p_k - \bar{p}) = p_k^1$ and $\bar{a}\bar{b} = \sum a_j b_j / N$,

$$W_j(p) = \left[\frac{N}{N-1} \cdot \frac{\overline{x_j^1 p^1}}{\sigma(x_j)\sigma(p)}\right]^2 = \frac{(\overline{x_j^1 p^1})^2}{m2(x_j)\,m2(p)} \tag{4.157}$$

4.8.6. Recognition with Supervision

A conceptual simple method for recognition with supervision is that based on nearest neighbors. Here the class to which a particular compound belongs is estimated from the class of its nearest neighbors. Here nearest is defined from the distance matrix, and in case of more neighbors the majority wins. The number of neighbors taken into account is found from statistical considerations and depends on the acceptable risk for misclassification and the number of members in the training set.

Another method is the development of linear pattern classifiers by means of negative feedback procedures. In order to apply these methods a training set should be available encompassing all classes and with about equal population in all classes. The aim now is the construction of a number of planes that each subdivide all patterns into two groups. The decision planes thus developed should all pass through the origin of the set of coordinate axes. In order to reach this goal, the number of coordinates that describes the position of a pattern in the R^n space is extended with one extra dimension (mostly chosen equal to 1). Now the orientation of a particular decision plane is fixed by a vector $\mathbf{w}$, perpendicular to this plane, at the origin. Multiplication of the vector $\mathbf{w}$ by the pattern vector $\mathbf{y}$ produces a scalar $s = \mathbf{w} \cdot \mathbf{y} = \mathbf{w}\mathbf{y}$ cos Θ in which Θ denotes the angle between $\mathbf{w}$ and $\mathbf{y}$. Depending on Θ, the cosine function is positive or negative. The sign of the s is used to develop the orientation of the decision plane: each member of the training set is multiplied by the decision vector $\mathbf{w}$, and each time a pattern is incorrectly classified the orientation of the decision plane and thus that of $\mathbf{w}$ is altered, for example, by a reflection operation in such a way that the distance between the pattern and the plane remains constant.

In situations where a linear separation between patterns is possible, the consecutive alterations of $\mathbf{w}$ should diminish and converge to a final value for $\mathbf{w}$. Sometimes, however, the patterns are not linearly separable and no convergence is achieved. Here the calculations, mostly performed with computer facilities, should be disrupted.

The procedure should be repeated for each set of two classes out of the training set. The set of decision vectors thus developed can be used to classify unknown patterns. The reliability of the process increases with an increase of the number of patterns in the training set (see Table 4.27).

Instead of negative feedback one may also apply a least-squares procedure to train the decision vectors. Here the difference between the actual s value s^* and the one calculated according to $\mathbf{w} \cdot \mathbf{y}$ is minimized for all y values as a function of all $n + 1$ elements of $\mathbf{w}(n =$ dimensionality of the problem = number of features):

$$s_i = \mathbf{w}\mathbf{y}_i = w_1 y_{i,1} + w_2 y_{i,2} + \cdots + w_{n+1} + y_{n+1} \qquad (4.158)$$

The error vector between s_i^* and s_i to be minimized,

$$R = \sum_{i=1}^{N} (s_i - s_i^*)^2 \qquad (4.159)$$

Table 4.27. Mathematical Procedure for Training a Decision Vector According to a Negative Feedback Procedure

1. Calculate $s = \mathbf{w} \cdot \mathbf{y}_a$:   (a) sign wrong, $\mathbf{w} = \mathbf{w}_{old}$; $s = s_{old}$
                                   (b) sign right, take new y value

2. Calculate $\mathbf{w}_{new} = \mathbf{w}_{old} + c\mathbf{y}_a$ with $c = \dfrac{-2s_{old}}{y_a \cdot y_a}$

3. Calculate $s_{new} = \mathbf{w}_{new} \cdot \mathbf{y}_a$

$$= \mathbf{w}_{old}\mathbf{y}_a - \frac{2s_{old} \cdot \mathbf{y}_a \cdot \mathbf{y}_a}{y_a y_a}$$

$$= s_{old} - 2s_{old}$$

$$= -s_{old}$$

4. Repeat step 1 with a new y value

5. Stop the calculation when corrections are no longer required for all y values involved

where N = number of patterns in the training set, yields $n + 1$ equations.

$$\frac{\partial R}{\partial W_k} = 2 \sum_{i=1}^{N} \left[\left(\sum_{j=1}^{n+1} w_j y_{i,j} - s_i^* \right) y_{i,k} \right] = 0 \qquad (4.160)$$

These $n + 1$ equations provide $n + 1$ values for the elements w_j of $\mathbf{w}$. The method with negative feedback as well as the least-squares method can be used only when the patterns are linearly separable, in contrast to the nearest neighbor method.

4.8.7. Classification with Neural Networks

Pattern recognition is possible in a quite different way, using neural networks. Neural networks are computer realizations of a model of the (human) brain. In 1943 McCulloch and Pitts (MC 43) postulated a model of nerve action that depended on the interaction of neurons. With this model they could explain some peculiarities in the way frogs handled visual information. McCulloch also stated that the neural model could be simulated in a Turing machine, the theoretical precursor of the computer. The availability of real computers made the development of a working model of neural action possible in about 1957 (RO 62). This "perceptron," as it was called, stimulated research in the field, but a very critical theoretical treatment by Minsky and Papert in 1969 (MI 69) drove all research underground. It was not until about 1985 that new developments were published, but since then research has been increasing at a very fast pace. Most research is in the psychobiological field: trying to understand the way the brain works. Part of the research aims at practical applications in control (of processes and movements), image and speech processing, and pattern recognition.

Neurons are cells in the nerve system. They are connected to many other neurons, and they are supposed to act only if they get a combination of signals from other neurons or detector cells that are above a certain threshold level. The computer model is built in the same fashion. A number of "neurons" act as input units. These units are connected to output units via a "hidden" layer of intermediate cells. All units are independent; there is no hierarchy. The network of connections between the units in the input layer, the hidden layer(s), and the output layer brings structure to the system. These networks can be altered by the user or even automatically or autonomously.

A unit works as follows:

1. Incoming signals (from other units or the outside world) are weighted with a factor.
2. The signals are combined to the net unit input. Usually, this is a summation of all incoming signals, multiplied by the weight factor. A threshold can be built in.
3. The new activity of the unit is some function of the net input and the old activity.
4. The output signal is some function of the activity of the unit. It can be a sigmoid function, a threshold, or some statistically altered function.
5. The output signal acts as the input signal for the units that are connected with the unit that delivers the signal. The signal can also act as an output signal for the output units and give a connection to the outside world.

In this system the weights can be altered by training. The activities and weights in fact contain the knowledge of the system. Training can be performed by applying certain rules to the adaptation of the weights when a given input expects a certain output. These rules are simple and stable and do not depend on the status of the system. The adaptation rules force the system iteratively to a stable situation. The whole system is thus defined by the number of units, the connection network, activity rules and weight adaptation rules and is independent of the problem that is offered to the input units, once the system has been defined. Training of the system with problems and their solutions is therefore independent of the connection between problem and solution. This connection can be totally heuristic. Problems that could be solved algorithmically can be offered also, but usually an algorithmic solution is faster. After the training is (nearly) completed (for instance, when the difference between the expected output and the learned output becomes stable) the weighing process can be frozen. Application of a problem then results in the learned output. As the activities of the units are simple functions, the output response can be very fast. Restarting the training is possible, though at the expense of more time. However, the training action, the adaptation of the weights, is a parallel action. Newer computers that work in parallel mode can make the training much faster.

Applications for neural networks can be divided into four groups:

1. *Pattern recognition*—Contrary to statistical pattern recognition, neural networks can handle symbolic knowledge. They are fast, and the training is easy.

2. *Memory*—Neural networks act in fact as nonorganized memories. After a successful training every trained input can be recovered without use of a systematic recovery system. Therefore the access time is short. The network is relatively insensitive to noise.

3. *Generalization*—When a problem that has not been included in the training set is presented to the trained network, the network always will provide an answer. When the answer is an interpolation between the problems used for the training, the result is probably quite good. The system is insensitive to exact matching. Extrapolation also gives answers, but here the results are usually

worse; only nice linear extrapolations give good results, but in that case extrapolations by algorithmic calculations are more precise.

4. *Processing of incomplete data*—The property of generalization can be used to compensate for missing data in the problem set. This property, combined with the possibility of continuous training, makes the system an ideal instrument for on-line decisions, e.g., in robotics.

In analytical chemistry there are not yet many applications. Some publications deal with the properties of the neural networks just described in items 2–4. The few publications using neural networks for pattern recognition are focused on supervised pattern recognition.

Classification of algae using some optical parameters (for instance, reflection and fluorescence at different wavelengths) obtained in a flow cytometer has been very successful (FR 89, SM 92). Long et al. (LO 91) have described an application on the classification of jet fuels. Other recent applications have been in the field of infrared (IR) spectrum interpretation (RO 90, SM 93). An overview of the state of the art in 1991 has been given by Zupan and Gasteiger (ZU 91).

4.9. OPTIMIZATION

Optimization of a (chemical) system requires an adaptation of all control variables in such a manner that the best possible result is obtained. This adaptation should be within the constraints set for the control variables of the system. A mathematical formulation of this situation could be to optimize a function φ that depends on n adaptable parameters X_i and m nonadaptable parameters Y_j:

$$\varphi = f = (X_i, Y_j) \qquad (1 \leqslant i \leqslant n, \quad i \leqslant j \leqslant m) \qquad (4.161)$$

Mostly the effect of the nonadaptable parameters on φ is considered to be random, which means that the results of variations of Y_j on φ are considered to follow a Gaussian distribution function around an average value of 0. Moreover the effect of Y_i and Y_j is assumed to be independent for all i and j ($i \neq j$). These simplifying assumptions lead to

$$\varphi = f(X_i) \qquad (1 \leqslant i \leqslant n) \qquad (4.162)$$

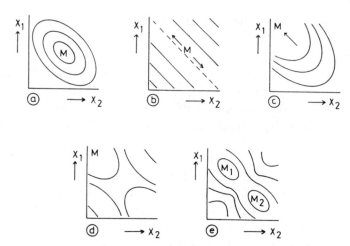

Figure 4.35. Two-dimensional representation of the shape of equi-response lines for various situations: (a) one discrete extremum; (b) a stationary rig; (c) a sloping rig; (d) a minimax saddle; (e) multiple maxima (or minima).

The graphic representation of the response of φ or variations of all x_i parameters is called a response plane. Connections between points with equal response on variation of various x_i parameters are called equi-response lines on the response plane. Five possible equi-response lines for a two-dimensional response plane are depicted in Figure 4.35.

Sometimes the relationship between φ and x_i is known as a mathematical equation with a set of boundary conditions. In these situations mathematical techniques for searching for optimal combinations of x_i values may be used, for example, in linear programming. When a mathematical relationship is unknown, experimental optimization techniques may be of help. Here a distinction can be made between simultaneous techniques and sequential ones.

4.9.1. Simultaneous Optimization Techniques

The initiation of this technique requires starting values for the x_i parameters under investigation and a set of combinations. The various combinations of parameter values produce a response plane. With two parameters a two-dimensional (flat) plane is formed; more parameters produce a more dimensional response space that cannot be represented in a flat plane without losing information. Some schematic examples of simultaneous optimization techniques are given.

1. *Random Design.* Suppose φ to be a function of two adaptable parameters x and y. A number of experiments can be designed for a set of combinations of x and y values selected at random. The experiment producing the best φ value is called the optimal condition. Another procedure based on this method starts from a set of increments for x and y values. The values $x_1, x_2, \ldots, x_a$ and $y_1, y_2, \ldots, y_b$ are combined in all possible pairs (x_i, y_j), with $1 \leqslant i \leqslant a$ and $1 \leqslant j \leqslant b$. From the set of all possible pairs a random selection of some pairs is made, and these are used as experimental conditions. This procedure is called *simple random design.* The location of the optimum can be found with enhanced accuracy by repeating the complete procedure in the direct neighborhood of the optimum found with a first trial. The operation of the random design procedure may be improved by application of some strategy for the selection of the parameter values that fix the experimental conditions. In the situation where two parameters should be selected, the response plane is subdivided into n equally shaped surface parts. Within each of these parts one randomly selected set of parameter values is used for the experiment. This method is called *strategic random design optimization*; it is also applicable for optimization problems with more than two parameters. In practice this method is most useful when a large number of parameters should be selected for a procedure that yields a small experimental error.

2. *Factorial Design.* The factorial design method for optimization starts by selecting an increment that is equal for all adaptable parameters $x_1, x_2, \ldots, x_a$ and $y_1, y_2, \ldots, y_b$. For example, x_1, x_2, x_3, and x_4 are combined with y_1, y_2, and y_3. Contrary to the random design method, all possible combinations (x, y) are investigated in order to find the optimal condition. Here too the accuracy is enhanced by a second trial around the optimal value with diminished incremental values. By duplication of the experiments, application of an analysis of variance scheme on the results is possible. This results in an estimate of the experimental errors involved and an estimate of the significance of the parameters under investigation. A disadvantage of this method is the large increase in the number of experiments for an increase in the number of parameters. This disadvantage may be partially circum-

vented by application of special strategies for parameter selection (half-factorial design).

4.9.2.   Sequential Optimization Techniques

An interesting possibility for experimental design is offered by a procedure that does not fix the selection procedure in advance. Here a start is made from a very limited set of experimental conditions. Based on the results of these experiments one (or some) more experimental condition(s) is (are) selected. Of course, the methods mentioned may be combined, and in practice various possibilities are applied. For instance, simultaneous method may give a global location of the optimum. Thereafter a sequential method may be applied for a more accurate location of the optimum. However, the inverse order may also used: in the neighborhood of an optimum found by a sequential method, a simultaneous optimization technique may be applied in order to verify the results and avoid false optima.

1. *Single-Step Procedures.* Here alternatively one factor is varied and all others are kept constant. The factor with the highest response, for example, y, is investigated first, followed by the next highest, for

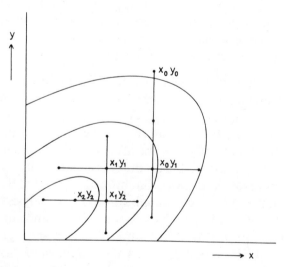

Figure 4.36. Order of measurements for a two-dimensional problem attacked with a single-step procedure.

example, x, and so on. A possible scheme for this method (Figure 4.36) is as follows:

a. Estimate the coordinates of the optimal condition; in the two-dimensional situation this could be (x_0, y_0).

b. From this origin x_0 is kept constant and four equidistant y values are selected covering the entire range of y values. An optimum is found for, say, y_1. Sometimes a better estimation is possible by fitting a polynomial function of order 3 through the four responses, which yields a better value for y_1.

c. A second set of experiments is carried out at y_1 and four equidistant x values covering the entire x range producing an optimal x_1 value.

d. A second cycle of optimization is started for both parameters, now searching with smaller increments.

e. The process is disrupted when, for instance, the change in the response is smaller than a predetermined limiting value.

2. *Steepest-Ascent Method.* Brooks (BR 59) gives two types of application of this method.

a. A "slope" procedure, starting from (x_0, y_0) according to a factorial scheme, say, (2×2): an estimation of the gradient based on the response of that scheme gives the direction of the steepest slope. Following this direction a new factorial scheme with interval S is constructed, producing a new estimation of the gradient. The procedure is repeated until the changes of the obtained response decrease. Finally a set of experiments is performed about the location with the optimal response, followed by a fitting procedure or a straightforward selection of the best response (Figure 4.37).

b. Single-step method: just as with the slope method, one starts with a factorial scheme around an estimated optimum (x_0, y_0) in order to find the direction of the highest gradient g_0. The second experiment starts at a distance S from (x_0, y_0) along g_0. The magnitude of the average response at this location is compared with that at (x_0, y_0): a higher response is followed by another experiment at a distance $2S$ from (x_0, y_0) along g_0, which is repeated until the average response stops growing (or until a boundary limit of one of the parameters is met). Hereafter a second direction of the gradient g_1 is calculated, followed by new experiments along this direction. The procedure is

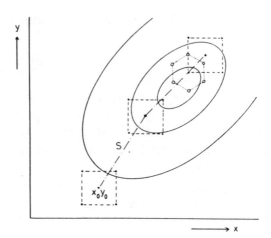

Figure 4.37. Steepest ascent method for a two-dimensional problem.

repeated until a disrupting criterion is met. The unique estimate of the optimal factor combination stems from the best response.

3. *Simplex Optimizations.* A simplex is a geometric construction fixed by a number of points equal to the number of dimensions of the problem space plus 1. The number of dimensions of the problem space equals the number of adaptable parameters, which in principle is not limited. The simplex method aims for a displacement of the simplex toward a region in the problem space with the optimal response. An equilateral triangle is applied for simplex optimization of a problem with two parameters. The angular points of the triangle are called vertices (DE 73). A tetrahedral simplex is applied for three parameters, and so forth.

As seen with the other optimization methods, usually only the most significant factors are selected for optimization. The relative importance of various factors follows from a factorial scheme. The importance of the factors may vary with the position in the problem space: around the starting location the order of importance can be quite different from that at the location of the optimum; therefore, at the end of a simplex optimization sometimes a fractional scheme is constructed in order to rank the influence of all factors on the response in the neighbourhood of the optimum.

The incremental distance for the factor variation is usually selected in such a way that the change in the response upon a stepwise change of a factor is about equal for all factors. At the initial stage a great increment is nearly always advantageous because of a quicker approach to the optimum. Moreover, under these conditions experimental errors play a less important role.

The boundary conditions for the factor space should be set in advance. These boundary conditions may result from limitation of temperature, pressure, solubility, and instability.

In order to position the first simplex, the starting values of all factors should be selected. In practice these values often result from previous experiments. The starting conditions do not affect the efficiency of the method in a significant way; however, if extra factors should be included at a later stage, it is advantageous to position one of the sides of the equilateral triangle parallel to the axis of one of the factors. Of course this is useful only when a visual inspection of a two-dimensional problem is made. Problems of higher dimensionality should preferably be handled with a computer. A detailed description of the order of the steps to be taken, simplex calculations, and warnings about pitfalls that may occur are found in a review article of Deming and Parker (DE 78).

A comparison of the simplex method with respect to other methods can be summarized as follows (Table 4.28):

Advantages
- A relatively small number of experiments to be made
- Simple calculations

Disadvantages
- No complete certainty about the absolute optimum—this also holds for other techniques; however, screening off by means of a suboptimum does not work for simultaneous techniques
- Handling a dimensionality greater than two rules out visual inspection and makes recognition of cycling around an optimal value hard to detect

Other situations arise when, apart from various analytical instruments, large computer facilities are available. Here networks may be constructed that feed the computer memory at regular time intervals with information produced by analytical instruments. Two different situations may exist.

Table 4.28. Comparison Between Simultaneous and Sequential Optimization Methods

Sequential Methods	Simultaneous Methods
A limited set of experiments is fixed in advance. Based on the responses of this first set of experiments the next set is designed.	A scheme for a great number of experiments is fixed in advance.
A response criterion (the density of the experiments in the experimentation space) is set in advance.	
After all experiments of the scheme are finished, the optimum is selected from the responses. Sometimes a second scheme is designed in the neighborhood of this optimum.	The series of experiments is disrupted when a criterion, set in advance, is met (this can be a minimal change in response during a fixed number of experiments).
Almost entire response plane is scanned.	Only a part of response plane is scanned, based on an estimated optimum and moving in the direction of the most advantageous response.
Erroneous high results of experiments easily result in false optima. To avoid this, experiments should be duplicated or a second scheme should be investigated.	Erroneous high results are detected because of their prolonged effect on a great number of consecutive experiments and the resulting optimization procedure. The result is corrected by duplication of the experiment.
Low accuracy of the optimum found.	Mostly the accuracy of the optimum is better.
Easily applicable experiments with prolonged duration.	Not easily applicable to experiments with prolonged duration.
Hardly any influence at all on the optimization of experimental errors.	The results are error prone.
False optima may be found when (too) high incremental values are applied.	False optima may be found with (too) big steps and with screening off.
	The technique may be automated. A graphic method can be performed without calculations (in the two-dimensional situations).

- The computer, as a master, asks each of the instruments at regular intervals for information. The instruments are "slaves" that produce information at the request of the computer.

- The instrument signals the computer that a set of measurements has been completed and will be transferred to the computer memory. Here the instruments sets the priority and acts as a master, using the computer as a slave.

Microprocessors and memory are cheap, so analytical instruments usually are fitted with their own private memories for temporary storage of a limited set of information until the master computer is ready to consume it. A general discussion of all possible situations is beyond the scope of this book, and special devices such as optical bar-code readers will not be discussed here.

ORGANIZATION

5.1. PLANNING

One of the quality aspects of analytical chemistry is the optimal execution of analytical tests with respect to time and money. As is shown in the preceding chapters much can be done by using the appropriate techniques for estimating the optimal sampling frequency, by selecting the optimal analytical method, and by applying optimal data handling. However, there are additional ways to influence the time, cost, and quality of analytical procedures. Much time and money can be wasted by improper planning of such aspects as availability of instruments, reagents, and people and improper routing, causing congestion and improper sequencing of actions. Much quality can be gained by having reliable forecasts of availability of new methods and instruments and estimations of the probability that new methods will be found.

All these factors can be included under the heading "planning." Planning in this context means arranging possible actions with respect to quality, quantity, and sequence. The basic objective of planning is to minimize time or cost and to maximize resource utilization. Planning thus sets the boundary conditions of a network of interrelated parts of the work to be done, the quantification of the size of the network links in time or money, and the reliability of the quantification, and planning also designates ways of selecting the optimal path.

Planning can be subdivided into forecasting, evaluation, and control. In this respect it resembles a common control problem, and in fact it should be considered as such.

In practice, however, many of the available methods are static and are used in a static way. A host of static and dynamic planning procedures can be found in the area of operations research (OR). In this chapter we discuss the concepts of OR from the point of view of analytical chemists.

5.1.1. Forecasting Methods

When one decides what work will or must be done in the future, there is a need for forecasting. Here one may be thinking in terms of some different time scales: the short, the medium, and the long term (Table 5.1). In the short run, at least, one is heavily constrained by an existing work load, but even here there can be an element of choice. Typically the work input exceeds one's capacity; thus a selection of tasks has to be made and, within that selection, an order of priority assigned. As the time scale increases, one's freedom of choice increases, at least in analytical work. One's choice is generally maximal in the longer term, and it is here that some of the forecasting techniques may be applied. There is a variety of approaches, but only a few that are applicable to analytical work and analytical laboratories are mentioned here.

"Good laboratory practice" (GLP) requires that in the laboratory a good organization exists, covering the responsibilities of the different stages of the analysis, and good records of what type of responsibility everybody can bear, by education, experience, etc. As we have done throughout the book, we do not give a description of how to organize a laboratory but rather a number of tools as well as hints as to how to build a flexible organization that can make quality analyses.

Table 5.1. Classification of Decision Problems in the Analytical Laboratory After Mitra's Extension of Anthony's Framework (MI 86)

	Objectives	Time Horizon	Nature of Information	Degree of Uncertainly
Strategic planning	Resource acquisition	Long, ~ 5 years	Highly aggregate	High
Tactical planning	Resource utilization	Medium, 0.5–1 year	Moderately aggregate	Moderate
Operations control	Optimal lab execution	Month, week, day	Detailed	Low
Analytical control	Optimal chemical execution	Day, hour	Detailed	Low

5.1.1.1. Trend Extrapolation

One assumes that the advances in a particular field will follow a defined pattern or progress in a relatively orderly manner, the so-called surprise-free projection. As a rule this extrapolation, for example, analysis time, analysis cost, and limit of detection, is performed in a quantitative way. With the aid of curve fitting (Section 4.4.1) the descriptor under consideration is described by an equation that fits the available data best. When the equation has been obtained, it is quite easy, although very dangerous, to extrapolate into the future.

The development of many analytical techniques during the last decades has been such that an exponential curve seems to fit the data,

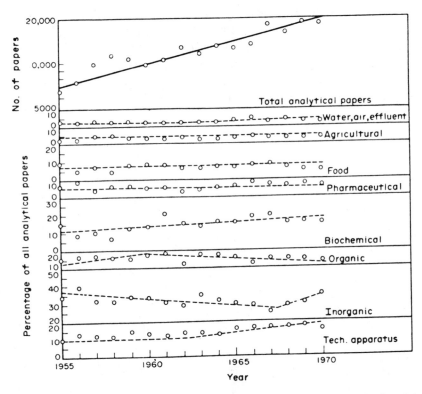

Figure 5.1. Development of analytical chemistry showing broad trends for eight categories of analytical chemistry for the period 1955–1970. Data expressed as percentage of total number of analytical papers. Reprinted with permission from Brooks and Smythe (BR 75). Copyright 1975 Pergamon Press Ltd.

for example, for the number of publications on analytical techniques (Figure 5.1) (BR 75, BR 76, BR 80). Maybe this is true, but not for the extrapolation. A more realistic curve for technical development is an S-shaped curve that levels off, resembling the first part of the well-known growth curve of microorganisms. However, the prediction curve of a particular technique can be connected to a curve of another newly developed technique that surpasses the former one. An illustration is the series of fitted curves on the number of papers devoted to separation methods (Figure 5.2).

A rough estimation of the speed or precision of subsequent colorimetric and spectrophotometric techniques will probably show the same trend. Separation methods such as analytical distillation, gas chromatography (GC), and high-performance liquid chromatography (HPLC) are other examples. For short-term extrapolation the method can be very useful, allowing estimations of, say, future work load, and thereby allowing predictions of the time when the use of automatic analyzers will be feasible. Here decreasing cost per analysis with an increase in work load allows estimations of cost in the future, when combined with trend analysis of labor cost.

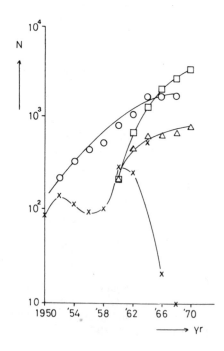

Figure 5.2. Development of analytical methods: (x) analytical distillation; (○) cumulative analytical distillation; (△) gas chromatography; (□) cumulative gas chromatography. Reprinted with permission from Brooks and Smythe (BR 75). Copyright 1975 Pergamon Press Ltd.

Another trend extrapolation method is time series analysis combined with predictive filtering methods, for example, Kalman filtering (see Section 4.3.2, esp. 4.3.2.2). This is a typical short-term method that has its merits in describing and forecasting analytical quality parameters, in particular when these are influenced by more or less random factors, for example, complaints from customers.

5.1.1.2. Needs Analysis

With this method one examines the needs expected in the future and estimates the probability that the necessary technology or expertise will be available in time. One of the applicable methods is the writing of a so-called surprise-free scenario. Here one describes all possibilities in a logical and justified way, estimating the probability of the possible alternatives. The "relevance tree" is pursued until all possibilities above a certain probability level for a predetermined time span have been described.

This method, which has been described in detail by Kahn and Wiener (KA 67), has been developed for such important forecasts as the economic development of countries. It is also used in a rudimentary form by many managers to explain why they need more instruments or equipment in the future.

5.1.1.3. Dynamic Modeling

Dynamic modeling comprises the use of mathematical models, analog or digital, to simulate situations and interactions expected in the future. These models are now being employed extensively in economic forecasting. One of the early models has gained much attention as the "world model" (FO 71), revealing the future situation in world economics, population, and pollution, for example.

Until now only a few examples of dynamic models in analytical chemistry have been described. Schmidt (SC 77b), Vaananen et al. (VA 74), and Doerffel (DO 75) applied digital simulation and queuing theory to analytical laboratory systems. Van den Akker and Kateman (AK 76) described a simulation model of an analytical laboratory to be used as a simulation game in the education of analytical chemists. Vandeginste (VA 77b, VA 79, VA 80, VO 81), Janse (JA 83, JA 84), and Klaessens (KL 88, KL 89) described models of a spectroscopic laboratory in a research institute and routine laboratories.

Because the delay of samples in an analytical laboratory usually exceeds the analysis time, most of the time is spent waiting. Here queuing theory may help to enhance some of the quality parameters of the laboratory, notably analysis delay time and cost. Vandeginste, restricting himself to a spectroscopic laboratory, stated that a dynamic planning strategy should answer the following questions (VA 77b, VA 79):

- Which strategy (or decision rules) should be followed to select the analytical method for solving the structure of the compound, taking into account the estimated probability that the analytical methods might solve the problem and minimize the queue lengths in the laboratory?
- What should be the procedure when an analytical method fails to solve the structure of the compound under investigation? After what analysis time should the analysis be terminated and another method tried?
- Which priority rules should be applied, and how do they affect the waiting time? (For example, what priority rules apply between samples of various origin, between samples that were unsuccessfully analyzed, and between easy and difficult samples?)
- Which kind of priority rule (absolute or relative) should be applied?
- What is the effect of interruptions of the analytical procedure by other activities on the waiting time?
- What is the influence of the structure of the organization, for example, centralized versus decentralized?
- Is it advantageous to accept samples only at discrete intervals?

Queuing theory answers these questions for simplified models only. However, from theoretical calculations on simple systems, the effect of the variables mentioned may be estimated for more complex systems. Vandeginste mentions the following results. The utilization factor (ρ) of the service channel plays an important role in all kinds of service systems, and an analytical laboratory is such a system. It is defined as

$$\rho = \frac{\overline{AT}}{IAT \cdot m} \tag{5.1}$$

where $\quad \overline{\text{AT}}$ = mean analysis time

$\quad\quad\quad \overline{\text{IAT}}$ = mean interarrival time

$\quad\quad\quad m$ = number of analysts serving the channel

For simple systems (the so-called M/M/1 systems[1]) where the samples are analyzed in the sequence of arrival at the system [the "first in, first out" (FIFO) rule], the mean waiting time is given by

$$\bar{W} = \frac{\overline{\text{AT}^2}}{2(1 - \rho)\overline{\text{IAT}}} \tag{5.2}$$

where $\overline{\text{AT}^2}$ = second moment of the analysis time.

The asymptotic shape of Figure 5.3 is characteristic for all queuing systems. From this figure it is seen that, for values of $\rho > 0.85$, minor variations in the organization of activities in the laboratory may induce a severe change in the waiting time.

It can be shown that decreasing the coefficient of variation of the analysis time—by processing many samples of the same kind for instance—decreases the waiting time, although the effect is less than the effect of a decrease of the mean analysis time, for example, by speeding up analyses. Figure 5.4 demonstrates that truncating the analysis and establishing a certain ratio between maximum analysis time and mean analysis time can improve the waiting time considerably if other channels are free. Otherwise the reverse effect may manifest itself. The effect of the mean interruption time β and the mean time between interruptions α (e.g., other activities while samples are waiting or breakdown of instruments) is given by

$$R = \frac{1 - \rho}{\alpha/(\alpha + \beta) - \rho} \tag{5.3}$$

where R = ratio of mean waiting times with and without interruptions; α and β are constants. This effect is depicted in Figure 5.5.

---

[1] In an A/B/C system, A depicts the type of arrival process, B depicts the type of service times, and C depicts the number of service channels (analysts); M means Markovian—each event is independent of the others; D means deterministic—each event is predictable.

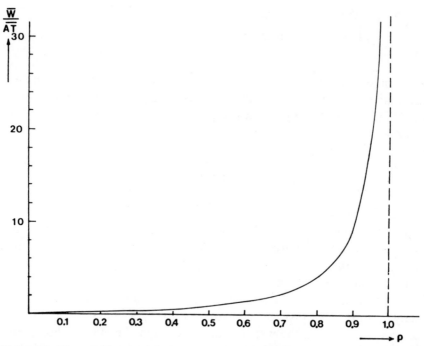

Figure 5.3. The ratio between the average waiting time ($\overline{W}$) and the average analysis time ($\overline{AT}$) as a function of the utilization factor (ρ) for a system with exponentially distributed interarrival times and analysis time (M/M/1 system) (VA 79).

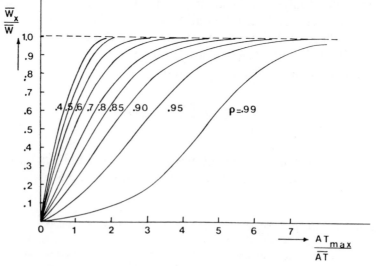

Figure 5.4. Reduction of the average waiting time, as a function of the maximal allowed analysis time $(AT)_{max}$. (VA 79).

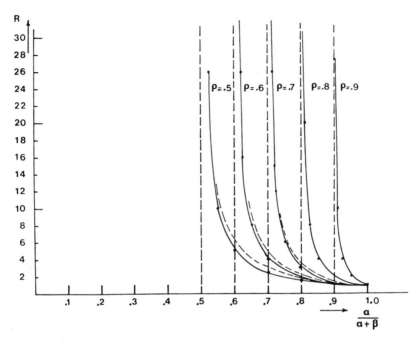

Figure 5.5. The ratio (R) between the average delay time for interrupted and uninterrupted analyses, as a function of the available time for analysis (VA 79): (—) α and β exponentially distributed; (---) α and β constants.

The priority given to samples can be effected in several ways. The effect of the various rules can be approximated. For priority groups with different mean waiting times the overall lowest mean waiting time is found when samples with the shortest analysis time are given absolute priority. Figure 5.6 shows the effect of this discipline as a function of the ratio of analysis times and number of samples in both classes.

In the analytical laboratory, this situation is met when an analyst executes two different analyses, or when the samples can be subdivided into two groups, for example, "easy" and "difficult" sample with "small" and "large" analysis times, respectively. Often, however, it is not possible to predict whether, say, a spectrum will be difficult to interpret or not. Thus after a certain interpretation time, the difficult spectrum may be transferred to a pile of unfinished spectra and the measurement of the next sample or interpretation of the next spectrum started. This problem can be solved only by simulation. In spectroscopic analysis, Vandeginste proved that waiting time can be improved

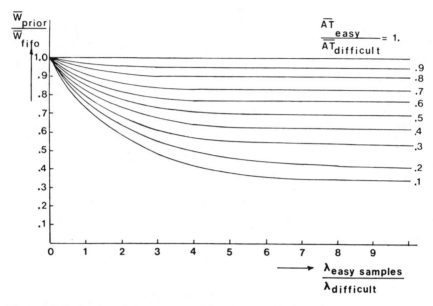

Figure 5.6. Reduction of the average waiting time, attributing absolute priority to the "easy" samples as a function of the ratio of the analysis time and ratio of the input density of "easy" and "difficult" samples (VA 79).

considerably only by giving priority to the "easy" samples that have been recognized as such beforehand. The results of the application of queuing theory on analytical laboratories can help in establishing rules for optimizing waiting time and cost in laboratories. However, in complex systems such as analytical laboratories, consisting of multiserver nodes and governed by state-dependent decision rules, the effect of the variables mentioned can only be calculated from simulations with a model of the laboratory under consideration.

Janse and Kateman (JA 84) simulated a routine laboratory. Some interesting results are shown for two laboratories in Figures 5.7 and 5.8. Increasing the sample load does not immediately increase the delay time. The delay time, an important quality factor, is defined as the time

Figure 5.8. Simulations: Effect of increasing sample load on mean delay time (D_t), mean waiting time (W_t), and mean idle time (dashed line). The vertical scale is in hours/day for the idle time: (left) for various sections (AN "0"–"5"); (right) for procedures with different priorities (JA 84).

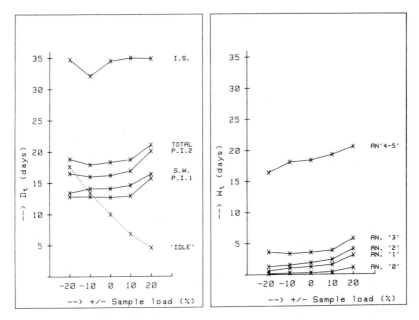

Figure 5.7. Effect of workload on delay times (D_t), per sections (P. I. 1 and 2) (left) and on waiting times (W_t) per priority class analyses (right).

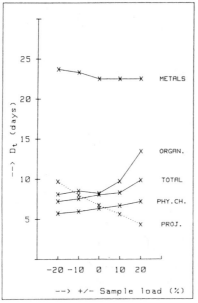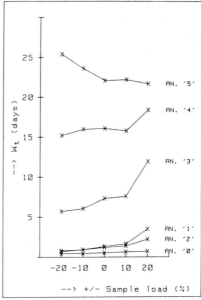

between the moment of sampling and the availability of the result of the analysis. Three main effects are responsible for this effect:

- High-priority work is effectuated; only low-priority work has to wait.
- "Idle" time [time used for (small) research projects, development and adaption of analytical methods, and recalibration] shifts to smaller jobs.
- The laboratory crew shows some "elasticity"; for a short period people can be motivated to work harder.

Combination of sampling theory (Section 2.6.4), information theory (Section 4.2), and queuing theory (Section 5.1.1.3) was possible for very simple laboratory models (only one serving point).

Figures 5.9, 5.10, and 5.11 show that the information that can be obtained from a laboratory used for control (a manufacturing process, a patient) depends heavily on the utilization factor (JA 83). The utilization factor is defined as the fraction of the theoretical laboratory capacity that has been used. When the total capacity of the laboratory has been used, delay is inevitable and the information output decreases. Depending on the properties of sample stream (frequency, distribution) and analysis type, the maximum information output is between 50% and 80% of the maximum capacity of the laboratory. It can be shown

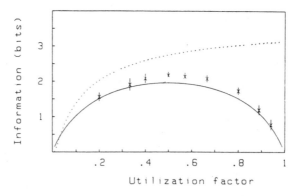

Figure 5.9. Theoretical curves for information yield as a function of utilization factor (JA 84): (—) M/M/1 system; (···) D/D/1 system, (×) simulation results with 95% probability intervals.

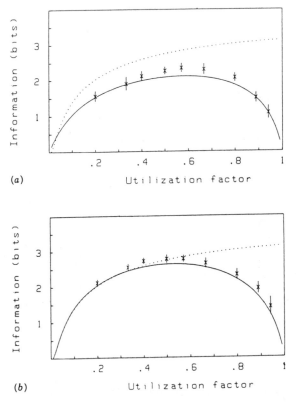

Figure 5.10. Theoretical curves for information yield as a function of utilization factor (JA 83): (a) (—) M/D/1 system; (⋯) D/D/1 system; (×) simulation results with 95% probability intervals; (b) (—) D/M/1 system; (⋯) D/D/1 system; (×) simulation results with 95% probability intervals.

that a D/M/1 system (see footnote 1) has a higher information output than a M/D/1 system. Making the sample stream more deterministic (predictable, better organized) results in a better output than organizing the analytical methodology.

An important, though not unexpected, result is that "last in, first out" (LIFO) gives a better information output than "first in, first out" (FIFO), as delay of the results can be diminished by discarding older samples.

Klaessens (KL 89) developed an automated simulation program for different types of laboratory. He showed that decreasing delay times was possible by optimizing a laboratory organization through a com-

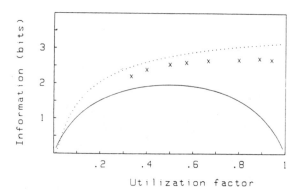

Figure 5.11. M/M/1 simulation results (JA 83): (×) with LIFO priority; (—) theoretical FIFO curve; (···) D/D/1 system.

puter model of that laboratory. Adjusting the number and type of instruments, the number of analysts, priority rules, and batch organization could result in lower delay times or higher information output at lower cost. His conclusions about delay time as a function of sample input confirm the conclusions of Janse.

5.1.1.4. Intuitive Modeling

Here one attempts to assess and refine the informed opinion of one or more experts. Although the intuitive approach is applied by most people, its more formal approach has not found widespread use in analytical chemistry. One of the methods available is the Delphi technique (DA 63). This technique depends on the individual views of experts on future developments in a particular field. The reasoning behind this method is that, apart from the better position of experts to forecast future developments, it may also be assumed that when the right experts are asked to participate they may shape the future by developing or by checking developments.

The Delphi technique applies a carefully designed program of sequential individual interrogations, interspersed with information and opinion feedback derived by computed consensus from the earlier parts of the program. Some of the questions may, for instance, pertain to the "reasons" for previously expressed opinions. A collection of such reasons may be presented afterward to each respondent in the group, together with an invitation to reconsider and possibly revise his/her

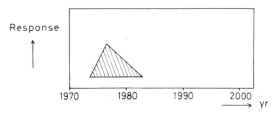

Figure 5.12. Results of a Delphi investigation in 1968 (HA 69) on an initial question: "After what year will large-scale information retrieval for science be available?" The corners of the triangle indicate first quartile, median, and third quartile.

earlier estimates. Both the inquiry into the reasons and the subsequent feedback of the reasons adduced by others may serve to stimulate the experts to reconsider features they may have inadvertently neglected and to give due weight to factors they were inclined to dismiss as unimportant on first thought. The consensus opinions, refined by this iterative process, are often produced in graphic form. In Figure 5.12 only a small part of a much larger Delphi exercise reported by Hall is shown (HA 69). The shaded area indicates the quartiles and median of the answers.

5.1.2.   Evaluation Methods

Once a reliable estimate of the future has been obtained or a set of alternative approaches to a problem is available in one way or another, it is desirable to evaluate the alternatives. An ideal way of planning—at least the most aesthetically pleasing way—is to pursue all possibilities. In practice, however, it is more realistic to drop the least promising ways and focus on the best. In such evaluations, the application of a decision tree may be helpful. Goulden (GO 74) presented an example of a decision tree shown in Figure 5.13. The example starts from the three options open to the analyst at the first decision stage. Success or failure of the options selected, together with the competitor activity (if any), will subsequently determine the options that stem from the second decision stage, and so on. One may also extend the tree with the probability of success in order to provide a measure of quantification that can be of further assistance in the decision making.

Massart (MA 75, MA 77) has demonstrated the use of operations research (OR) techniques in analytical chemistry for evaluating alternatives in analytical procedures.

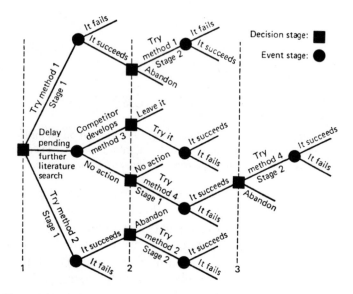

Figure 5.13. Decision tree for analytical method development (GO 74).

5.1.3. Control

The final stage in planning is planning control. This can be achieved in numerous ways, but only a few are mentioned here.

5.1.3.1. Checklists

The most primitive but useful aids in control of planning are checklists. One prepares a list of actions that should be performed before the ultimate goal will be reached. During the procedure, regular consultation of the checklist prevents deviations from the desired way. These checklists can be made more or less complex by introducing order, for example, arranging actions in the sequence that is optimal or in order of importance to the final result.

5.1.3.2. Gantt Charts

A checklist in graphic form is the Gantt chart, the well-known bar representation of activities on a time scale. The graph shows a number

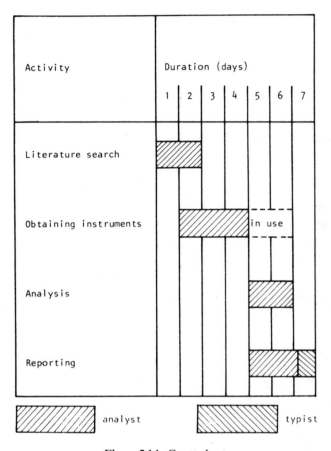

Figure 5.14. Gantt chart.

of activities, the anticipated date of completion of these activities, and the means to be used (Figure 5.14).

As is shown in the example, the chart not only displays the check-list, in this example a very short one, but also gives insight into the occupation of instruments and the need of external services, for example, typists. Many layouts of such charts are available through firms that supply managers with planning aids. The host of gadgets, multicolored plastic cards, vast planning boards, and computer pro-grams tends to obscure the usefulness of this method. Particularly in the more routine areas, where the time required for the successful completion of individual tasks can be predicted with fair certainty, the Gantt chart method can aid in the control of complex operations.

5.1.3.3. *Network Diagrams*

A control system more advanced than the bar chart is the network diagram. Two main types have been developed, PERT (Program Evaluation and Review Technique) and CPM (Critical Path Method). In essence their principles are alike. The planning of a project with a network can be divided into three stages:

1. Subdividing the project into a number of separate unit operations and rearrangement into a network.
2. Estimating the time required for each unit operation and setting the critical path—the shortest possible way—through the network.
3. Pinpointing the critical points and delays and taking precautions to avoid or diminish these obstacles.

The basic rules for establishing a planning network are quite simple. The planning network can be drawn and constructed by simple arithmetic. For more complicated jobs, and the by then very complicated network, a number of computer programs are available that perform the same tasks. Most computer programs provide the user with plotted outputs.

The procedure for establishing the network is as follows:

- Each unit operation or action is represented by an arrow. The length of the arrow is not important.
- The start and completion of an action are indicated by numbered circles called *junctions*.

Each action arrow and its junctions can be drawn when the following questions are answered:

- Which action precedes the one under consideration?
- Which action proceeds simultaneously with this one?
- Which action follows this one?
- What defines the completion?

A complete network has only one starting point and one terminal. Numbering of the junctions can be done as follows:

1. Find the junctions without incoming arrows (with only outgoing arrows).
2. Number these consecutively in the order in which they are found.
3. Delete all arrows that depart from numbered junctions.
4. Apply rules 1, 2, 3, etc.

The network can be completed by establishing the time of completion each action. Now it is possible to indicate at each junction the earliest possible time of completion by adding to the preceding earliest possible time the time indicated by the incoming arrows. The highest value obtained is noted at the junction. The last allowable moments are obtained by working backward in the same way as used when estimating the earliest possible time. The difference of both times is the idle time for that junction. The path along the junctions without idle times is the critical path. This determines the shortest possible time for completion of the whole task.

Figure 5.15 shows a very simple network, and Figure 5.16 shows an alternative way of drawing this network, which allows a quick survey of all idle times and the critical path.

The expected time of completion of the whole task can be estimated by

$$t_E = (A + 4M + B)/6 \qquad (5.4)$$

where A = most optimistic estimation of time of completion
   M = most probable estimate of time of completion
   B = most pessimistic estimation of time of completion

The variance of this estimate is

$$V(t_E) = (B - A)^2/36 \qquad (5.5)$$

The standard deviation is then

$$\sqrt{V(t_E)} = \frac{|B - A|}{6} \qquad (5.6)$$

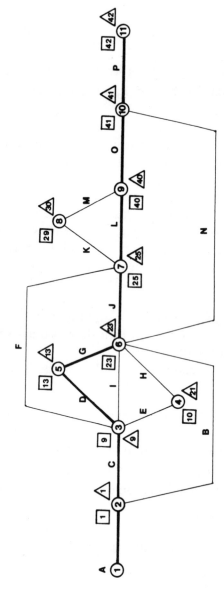

Figure 5.15. PERT network planning: A, sample arrives; B, preexamine sample; C, survey literature; D, order chemicals; E, select method; F, borrow instrument; G, read technical information; H, prepare method; I, wait for chemical; J, test method; K, prepare sample; L, order instruments; M, prepare test samples; N, wait for instruments; O, test instruments; P, measurement; (○) most likely time (days); (□) earliest possible moment; (△) latest possible moment; (—) critical path.

280

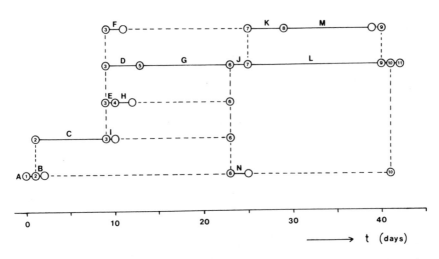

Figure 5.16. Planning network with linear time scale: (—) active time; (---) passive time. For explanation of symbols see Figure 5.15.

It must be emphasized that the assumptions used in deriving these equations are not generally accepted. However, the equations provide acceptable accuracy in practice. When a 3σ level is accepted for the probability that an estimate will occur in practice, then it is clear that the expected time of completion will have a value between A and B (as might be expected!).

5.1.3.4. Flow Charts

The extensive use and acceptance of flow charts in analysis and planning of computer programs has led to the development of the flow chart technique for planning in (analytical) research. Although Davies (DA 70) was the first to publish a proposal, the method seems to have been developed independently in several places. In the diagram actions are represented by rectangles, decision points by diamonds, interdependent points by circles, and end points (failure or success) by rounded rectangles. The diagram has the advantage that recycling and feedback sequences can be shown. These activities are an essential part of research activity. Quantification of probabilities and times for the activities and decisions is possible and allows the estimation of probabilities of failure or success and of time of completion. For the more

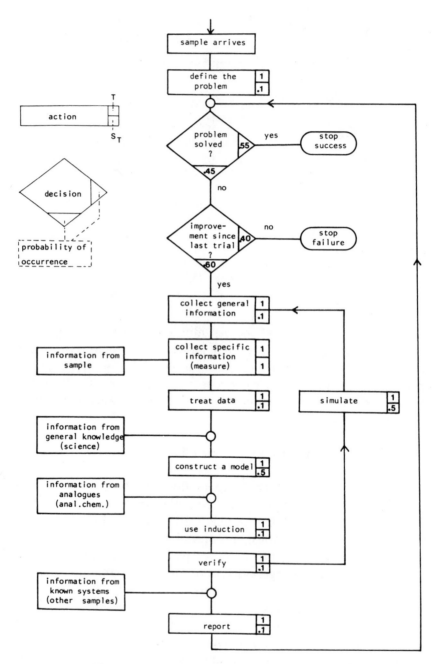

Figure 5.17. Flow sheet of analytical work (GO 72).

complicated jobs computer simulation is feasible. In Figure 5.17, a flow chart, derived from one first published by Gottschalk (GO 72), depicts an analysis of an unknown sample.

5.2. ORGANIZATION AND MANAGEMENT

Nowadays most organizations change, voluntary or not, because of a number of internal and external causes:

- Organizations expand and shrink; they are seldom of stable magnitude.
- Essential standards of society change.
- The influence of the external factors on organization increases.
- The pattern of moral values and standards of members of the organization changes.
- The social factor in organization increases.
- Technical factors in organizations may change dramatically (computer).

In summary, society as well as people change their standards and values, and organizations change within themselves and with respect to their relationships.

One may wonder in what respect analytical laboratories are influenced by the factors mentioned here. It is not feasible to elaborate on all psychological, social, and economic features. Moreover, it is impossible to select one particular system to operate a laboratory. In general the preference for one system over others depends on the institute or organization where one operates, and indeed the laboratory may contribute only partly to the final decision.

5.2.1. Description of an Organization

Various experts on organization theory have tried to give an adequate definition of organization. There are numerous definitions, and a selection out of the collection of necessity is subjective. The following definitions should be considered as an illustration of the possibilities focused on organization as a tool of management. The number of basic principles is also limited because of the selected definitions. These limitations imply that the term organization is not used for description of management and that organization is a tool, not a goal. Thus—

- Organization is a complex of relative permanent measures aiming for a well-operating assembly of means in order to achieve a goal set in advance.
- Organization is the ordering of actions and creation of an effective assembly of means in order to reach a goal.
- Organization is the creation of effective relations between people and means in order to achieve a goal.
- Organization is a social structure, constructed from coordinated positions referring to a common purpose.
- Organization is a distribution of labor among a group of co-operating people in order to make the company as good as possible with a minimal effort and the highest job satisfaction to all members of the group.
- Organization is the complete set of rules and standards designed to improve efficiency.
- Organization is a complex system consisting of necessary actions and alternative applicable means assembled and governed in order to achieve a given purpose. The system of actions and means interrelates the cooperation of, for example, human–human, human–group, and human–means.
- Organization is a frame of functions, tasks, and processes, embracing people that aim for a common goal with limited, alternatively applicable means.

The frame is set by the following:

1. Inventory of functions, processes, and procedures
2. Creation of bodies to operate in the functions
3. Effective subdivision of functions and tasks among persons and equipment, including a pattern of relations

An adequate description for our purpose will contain various features such as the following:

- Functions, tasks, rules, standards, and conditions
- Processes and procedures
- Cooperation interrelations, economic interrelations, and so on

These features may be described from three different points of view: technical, economic, and sociopsychological. Each of these areas has more or less its own stabilized organizational theory.

A technician focuses her/his attention on the operation, control, and ruling of processes and procedures. The economist formulates the boundary conditions for the technical systems. The social psychologist develops a system for the functioning, rule, and harmony of the working community and its relationships. She/he distinguishes, on the one hand, a social system for the operation and cooperation of the staff in order to achieve the goal of the company and, on the other hand, the private aims of the individual and the team. In such a social system one may distinguish the rules of the game one must obey in order to obtain a well-functioning process. Apart from this, one must stay within the boundary conditions set by economic and technical limitations. An economist considers the ranking and functioning of various alternatives for means of production such as labor, equipment, capital, and chemicals to such an extent that the company goal is obtained with as little effort as possible and/or that the yield is as high as possible and/or that the continuation of the institute (company) is warranted. Here the technical and social systems set the boundary limits. Each of these systems distinguishes conditions, and rules of the game that have to be obeyed in order to allow the functioning of the process. Moreover, each system sets limitations on the other two, defining the boundaries within which each may operate.

The social system formulates the conditions with respect to human power of endurance—physical as well as psychical—that is, working hours, informal group decisions, and so on. The economic system formulates conditions in minimal yields and budgets, and the technical system gives conditions on instruments, for example. As well as attention, knowledge of the question is involved apart from knowledge of investment and operating costs. The former conditions pertain to the social system, the latter to the economic system. In general, one may conclude that an organization should be considered as a complex system composed of at least technical, social, and economic subsystems that may be distinguished but cannot be separated from each other.

Now we come to the question of how to describe this system in a scheme that visualizes disturbances, changes, and so on. This scheme should also allow evaluation of the results of actions aiming for optimization. The normal schemes of organization with their treelike constructions are completely inadequate for this purposes; at most they describe a formal part of the personnel organization. The entire picture of the organization is not given. A scheme exists that partly serves these purposes; it is described below.

Figure 5.18 depicts an organization as a system of actions or tasks (i.e., actions not yet performed) and means with a mutual relationship. This scheme is based on the assumption that in each organization a set of fundamental actions and means exists and can be listed. A matrix

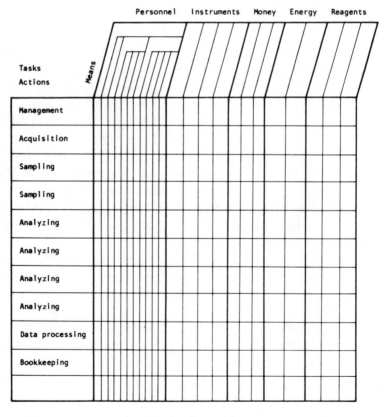

Figure 5.18. Organization scheme interrelating tasks and means as a function of time (SC 66).

representation of this scheme gives the means along one axis and the tasks along another. The mutual relationship between a task and the available means are represented by matrix elements. Figure 5.18 summarizes all procedures and processes required for good operational conditions. The system operates well, provided that each matrix element contains the right conditions. The conditions pertain to boundary limits and to basic requirements. Extension of the two-dimensional matrix with a third dimension gives an indication of the cost and prices per means per time unit. A technician will preferably focus his attention on the first two axes of the matrix; however, the entire problem covers all axes of Figure 5.18. The social system of the organization is found mainly within the column titled personnel. Here cooperation and harmonious relations between individuals and departments play a role. However, each matrix element may be denoted by incoming and outgoing information streams, pertaining to process and procedures and addresses and where they come from or go to, thus making the social system operational in the entire matrix. The economic system of the organization pertains to all means that carry a price and are consumed on reaching the goal. An important implication of the matrix representation is that each specialist who alters something in her/his subsystem invokes an action in the subsystems of all colleagues belonging to the organization: the three subsystems mentioned may be distinguished but are inseparable!

Even Figure 5.18 does not quite describe the entire problem, because the organization is a dynamic system of actions and means. Each action requires a period of time for it to become completed. The dimension time is invoked by attributing a time span to each matrix element. Because of internal and external forces, the representation matrix of the organization is changing continuously with respect to all dimensions.

5.2.2. Structure of an Organization

The method of representation as presented in the preceding subsection provides the necessary information on the interactions within an organization but not on its structure as it is or as it should be. Actually it is merely a structure of relationships. Other structures that may be mentioned are a communication structure and a structure of responsibilities, or a hierarchical structure. The structure of the communications can be derived from the structures of relationships, because communication means transfer of information. It is possible to visualize the information channels—their location or their desired location—by extracting from the relation scheme the part pertaining to information. One should keep in mind that information does not pertain exclusively to technical information. Social and economic information should be considered as well.

There are various methods with which to perform communications, the most important ones in a laboratory being the following:

1. *Electric signaling.* Here both emitter and receiver are interconnected with electric channels. A piece of information is coded and emitted and at the receiving end is decoded. The communication network is mostly centralized within instruments and is unimportant from an organizational point of view. Communication between instruments (e.g., a titrating unit and a computer) as a rule occurs at a level too detailed to be inserted in an organization scheme. At locations where the information stream branches off a further inspection should be made. Here it should be considered whether the branches are optimally positioned or may be omitted (e.g., an electric signal toward a recorder that is not operational).

It is also important to investigate the occurrence of an eventual shortcut in the information stream that may cause a lack of information at a location where it is needed, for example, an operator who does not get the information required for efficient operation of the instruments.

2. *Information storage.* An information channel that should be mentioned separately is a mechanical one toward paper, recorder strip charts, computer listings, and magnetic and optical disks. The main characteristic of this channel is its ability to bridge time spans. This property has an advantage as well as a disadvantage: it builds an archive for seconds up to centuries that may easily be reproduced.

Because of the considerable time delay that may be realized, it is sometimes hard to see whether the aim, that is, transfer of information, is reached. The structure of the communication lines, as obtained from the structure of the relations, should be demonstrated to see whether the paper communication serves its purpose, taking into account an eventual archivation. Here the following questions should be answered. Is the information required for, say, GLP (Good Laboratory Practice)? Is the information given to people who should not receive it? Is information filed for future use still up to date and useful when required? Is enough information available for interpretation when archives are inspected after a great time delay? Will stored information be inspected in the future? Can information be retrieved?

3. *Acoustic and optical communication.* This type of information transfer uses acoustic or optical data channels. In a laboratory the human emitter and receiver of acoustic signals are the most important ones. Optical emitters are found in instruments such as lamps, led displays, and analog meters such as clocks, and also in paper. Sometimes optical receivers are found inside instruments such as photodetectors; when they are found outside instruments; the human receiver is the most important one. Both types of communication are denoted by high-speed signal transfer, the possibility of handling complicated patterns of information such as pictures, music, and speech and their transient behavior.

4. *Communication using human beings.* It was mentioned above that the human being may act as an emitter as well as a receiver of optical and acoustic information. The human being may also serve as an archive for this information, which enables the bridging of time gaps, as do paper archives. However, important differences to be noted are that human memory is rather unreliable and within the human mind information can easily be garbled. These two features are probably interrelated. A result is the involuntary and even unknown falsification of information. An advantage of combination of internal communications in the human mind is the relatively easy decoding of complicated patterns of information. A human mind may simplify, interpret, and recognize patterns.

The judgment or construction of a communication structure within a laboratory should proceed according to the following rules:

- Derive the pattern of communication from the information part of the relation structures.
- In situations where a time gap should be bridged, paper or magnetic computer memory devices (tapes, disks) should be selected.
- In situations where information should be combined or interrupted or when pattern recognition should occur, human input should be applied, eventually in combination with computer techniques.

- All cases of straightforward data transfer should be handled along electric, optical, or acoustic channels. The last two ways are of special importance for communication with people and for inter-human communication. The other channels are of importance for interdevice communication.

- Communication serves information transfer; without a receiver a channel is useless!

Finally, a laboratory communication scheme may be presented in Figure 5.19. Here exclusively technical and economic communication lines are drawn. It must be possible to extend this scheme with a social structure. This could reveal some surprising facts. A bad technical communication is often caused by disturbances in the social com-

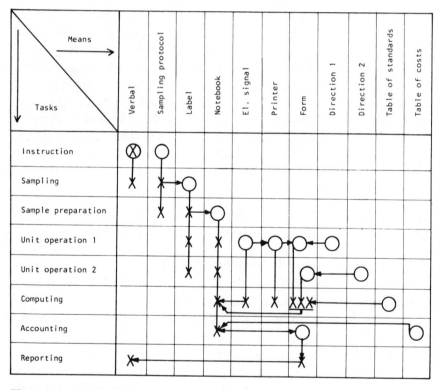

Figure 5.19. Relation between tasks and tools in analytical chemistry; tasks presented in chronological order.

munication channels: an unapproachable boss, people from hostile communities, or an immigrant worker who does not fully understand the language of the host country are some causes that may easily be detected. Beneficial influences may result from people such as secretaries or assistant managers if they are competent. They make the communication lines shorter and more efficient and operate with fewer mistakes. Research on this subject has revealed that these informal communication lines generally do not correspond with the official lines, which correspond with the responsibility structures (AL 77, PE 66).

Some other factors to be mentioned are as follows:

- *Geographic distance*—there seems to be a reciprocal exponential relationship between communication and distance.
- *Social "wavelength"*—language, habits, regional origin, etc.
- *Ability to interpret*—i.e., to process difficult information in a manner accessible to the receiver.

In Figure 5.20 models for analytical laboratories are given that focus more on the flow of information. Models that look like these are used for the description of a laboratory for simulation or for a so-called LIMS (laboratory information management system) (MD 87). In essence, a LIMS is a data base for storage of laboratory data. It has a number of modules, depending on the type of LIMS, each taking care of a specific part of the data management system.

A typical full list of modules looks like this:

- *Sample log-in*—This part takes care of the bookkeeping of the incoming workload and keeps in memory what should be analyzed in the sample. Also the origin of the sample, the name of the principal, etc. can be stored in this module.
- *Work-load management*—Performs activities such as showing the progress of the sample on generating worklists.
- *Data acquisition*—Data can be entered manually or by direct connection to the instruments. Data can be verified for the right format or the allowed range.
- *Report generation*—Most LIMSs have a facility to type a report with all relevant data in a predetermined format. All extra information can be added here.

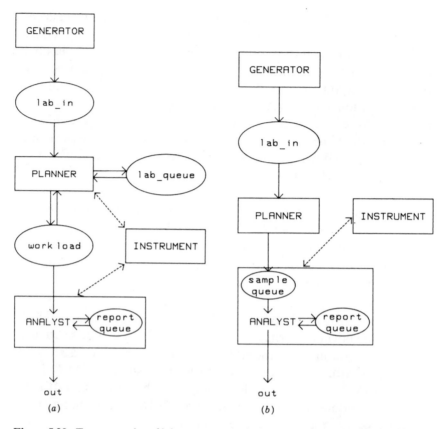

Figure 5.20. Two examples of laboratory organizations: (a) with centralized planning; (b) with decentralized planning. In boxes: laboratory components; in ellipses: queues for sample storage; full arrows: interaction (KL 88).

- *Archiving*—Keeping all raw data and results is obligatory for many quality systems, such as GLP. It is important to keep these data in such a way that they can be found easily when required. This seems to be the most difficult part of a LIMS, and many users prefer archiving on paper.

A LIMS is a very important instrument for quality control. In the first place all data are stored and can be retrieved, so quality loss through lost information, copy errors, etc. cannot occur. All users are obliged to follow the same rules, exceptions having to be marked as such. Any irregularity is discovered. A second important aspect is that most systems have some statistical treatment of the results, so accuracy, precision, and drift can be easily obtained. It must be remarked, however, that this aspect of the LIMS has not been exploited fully. The availability of all data concerning the analysis process should allow much more quality control.

A LIMS is a commercial product. Advantages are that training facilities are available and help at the introduction is usually part of the system. A disadvantage is that the results of the quality control module are not easily comparable to the results of other systems. Another drawback is that often it is not clear what algorithms are used for the quality control procedures. Recently research on the more fundamental aspects of LIMSs has started, resulting in a scientific journal dedicated to the subject (MA 91).

5.2.3. Structure of Responsibility

The hierarchical structure of a laboratory depends greatly on that of the entire company. The latter is established by historical, philosophical, and political considerations. Therefore it is not very useful to discuss laboratory organization in general. However, within the responsibility of the department one may have an opportunity to effect a change in the hierarchical structure, and so some remarks may be made here. Moreover in modern quality control, e.g., GLP, the hierarchical structure has to be such that it is always possible to trace who was responsible for certain results, and it is stressed that responsibilities are taken only by people who are trained to bear such responsibilities. So in many quality assurance systems a record should be kept of the training (date, subject, results) of the personnel. In view of the theories of organization discussed above, it seems desirable to delegate responsibilities, as far as possible, to persons or a group of people. The following criteria should apply:

1. The responsibility should be such that it can be successfully undertaken. The person under consideration should have a personality that can cope with the responsibility in terms of emotion, experience, and knowledge. In other words, one should have the ability to perform all actions that are necessary to do one's part of the job. In order to do this one must have the appropriate character and education, as well as the means to plan a budget, and wield decisive authority. One should

have the power to apply these means freely, to use them at will. A characteristic feature of the situation is that one must be accountable for all duties one is to perform but not for details as to how they are done.

2. Control of the combined duties should be possible. Apart from the means to perform controlling actions, one needs information to compare the results of actions with respect to standards. The structure of the organization should reveal whether the person under consideration may get the appropriate information. Important technical information refers to knowledge about the characteristics of procedures that are applied (such as accuracy or speed) and the standards that should be set (dictated by the process under consideration). Information on economics comprises expenses such as cost per analysis, fare composition, overhead, and charges for equipment and chemicals, on the one hand, and available budgets, on the other. Information on social circumstances comprises knowledge of personal merits, labor situations, job satisfaction, disturbances, and the like between interrelated people and people that interact with the group. Standards are found in labor rulings and other formal regulations, as well as in social and ethical norms.

The area that is covered by the responsibilities of a laboratory technician is set by a number of features partly belonging to the person and partly to the technician's supervisors. The positioning of the responsibility area within the structure of relationships is governed by various rules. Some demands are as follows:

- The responsibility should be bearable.
- The entire area of relationships should be covered by various responsibility areas.
- The structure of the communication network sets the organization (i.e., for an information-producing organization such as a laboratory) and not the hierarchical structure.
- In situations where an overlap exists, the communication line should point in one particular direction.

Nadkarni (NA 91) has given 10 rules for total quality management, including some on the subject covered here:

Commitment. Believe in the philosophy of quality and show constancy of purpose because it's the right thing to do; it is here to stay for the foreseeable future.

Leadership. Show quality behavior by example. Leadership is for everyone, not just managers.

Teamwork. We are all in this together—subgroups, departments, suppliers, chemists, engineers, service staff and research staff. Build on each others ideas and strengths.

Empowerment. As a boss, take the blame and pass on the credit. Give people power, responsibility, pride of workmanship, and recognition.

Pride of workmanship. Enjoy your work. However bleak things may look, you can hold your head high knowing that you did the job your way, that you did it the right way, and that you did it the quality way.

Some structures of responsibilities may be presented:

1. *Organization according to unit operations.* Each organizational unit—one person or a group—bears the responsibility for one action or a particular group of actions. The advantages of this type of organization are the minor requirements of knowledge, the ease of automation and mechanization, and the economic application of analysis on a routine base. A disadvantage is the high information transfer from person to person.

2. *Organization according to procedures.* Here each organizational unit is responsible for one procedure or a particular group of procedures. Quality control is of the utmost importance. Its advantage is mainly the minor information transfer between people; its disadvantages are the enhanced possibility of carryover, which here means that analytical results of different samples are interchanged by mistake, and the increased requirements of knowledge.

3. *Organization according to duties.* Each organizational unit is responsible for a particular area of tasks, for instance, the production of information about a particular plant or product or for a research group. The advantages here are a very limited transfer of information between the units and a clear responsibility for production or for the results of research. The disadvantages are the heavy demand for

knowledge and the possibility of a shortcut within the laboratory organization. Some of the communication channels become more or less artificial.

4. *Mixed organizations.* A situation often encountered in practice but not tested for optimal conditions with respect to organizational behavior is as follows: an organizational unit (person or group) is responsible for one task or a combination of tasks and for one procedure or a combination of procedures. Here the entire area is covered better and the lines of communication follow a natural pattern, which is advantageous. The disadvantages are the overlap in the organizational structure and the economic structure, which is hard to visualize. It is difficult to determine the costs of particular actions and to calculate the overhead on instruments. An excellent survey of management studies pertaining to analytical research is given by Goulden (GO 74), who describes the consequences resulting from the employment of various managerial styles.

REFERENCES

AC 83 ACS Committee on Environmental Improvement: *Anal. Chem.* **55**, 2210 (1983).

AG 71 R. B. d'Agostino: *Biometrika* **58**, 341 (1971).

AK 76 M. J. A. van den Akker, G. Kateman: *Fresenius' Z. Anal. Chem.* **282**, 97–103 (1976).

AL 77 T. J. Allen: *Managing the Flow of Technology.* MIT Press, Cambridge, MA, 1977.

AM 77 A. E. Ames, G. Szonyi: *ACS Symp. Ser.* **52**, 225–227 (1977).

AN 75 Analytical Chemistry: *Anal. Chem.* **47**, 2527 (1975).

AN 76 Analytical Chemistry, Division IUPAC: *Pure Appl. Chem.* **45**, 101 (1976)

AR 72 Arbeitskreis: "Automation in der Analyse": *Fresenius' Z. Anal. Chem.* **261**, 1 (1972).

AR 73 Arbeitskreis: "Automation in der Analyse": *Talanta* **20**, 811 (1973).

AS 71 American Society of Testing Materials: *Standard Recommended Practice for Use of the Terms Precision and Accuracy as Applied to Measurements of a Property of a Material*, ASTM Standard E 177-71. ASTM, Philadelphia, 1971.

AS 73 American Society of Testing Materials: *Sampling, Standards and Homogeneity*, ASTM STP 540. ASTM, Philadelphia, 1973.

BA 28 B. Baule, A. Benedetti-Pichler: *Fresenius' Z. Anal. Chem.* **74**, 442 (1928).

BE 47 W. P. Belk, F. W. Sunderman: *Am. J. Clin. Pathol.* **17**, 854 (1947).

BE 75 R. J. Bethea, B. S. Duran, T. L. Boullion: *Statistical Methods for Engineers and Scientists.* Dekker, New York, 1975.

BO 64 G. E. P. Box, D. R. Cox: *J. R. Stat. Soc.* **B26**, 211 (1964).

BO 70 G. E. P. Box, G. M. Jenkins: *Time Series Analysis (Forecasting and Control).* Holden-Day, San Francisco, 1970.

BO 73 B. Bobec, M. Lachance, O. Cazaillet: *Eau Que.* **16**, 39 (1973).

BO 78 P. W. J. H. Boumans: *Spectrochim. Acta* **33B**, 625 (1978).

BO 79   B. Bobec, D. Bluis: *J. Hydrol.* **44**, 17 (1979).

BO 83   J. Bouma: *Agric. Water Manage.* **6**, 177 (1983).

BO 85   J. Bouma: In *Soil Spatial Variability* (D. R. Nelson, J. Bouma, Eds.), p. 130. Soil Survey Institute, Wageningen, The Netherlands, 1985.

BR 59   S. H. Brooks: *Oper. Res.* **7**, 430 (1959).

BR 62   R. G. Brown: *Smoothing, Forecasting and Prediction of Discrete Time Series.* Prentice-Hall, New York, 1962.

BR 75   R. R. Brooks, L. E. Smythe: *Talanta* **22**, 495 (1975).

BR 76   T. Braun: *Talanta* **23**, 743 (1976).

BR 80   T. Braun, E. Bujdoso, W. S. Lyon: *Anal. Chem.* **52**, 617A (1980).

BR 84   G. Brands: *Fresenius' Z. Anal. Chem.* **314**, 646 (1984).

BR 86   S. D. Brown: *Anal. Chim. Acta* **181**, 1–26 (1986).

BR 89   J. P. Brans, B. Mareschal: *PROMETHEE-GAIA: Visual Interactive Modelling for Multicriteria Location Problems*, Internal Report University of Brussels, STOOTW/244, 1989.

BR 90   R. G. Brereton: *Chemometrics, Applications of Mathematics and Statistics to Laboratory Systems.* Ellis Horwood, New York, 1990.

CA 75   J. P. Cali: *The Role of Standard Reference Materials in Measurements Systems*, NBS Monogr. No. 148. U.S. Govt. Printing Office, Washington, DC, 1975.

CE 75   G. S. Cembrowski, J. O. Westgard, A. A. Eggert, E. C. Toren: *J. Clin. Chem.* **21**, 1396 (1975).

CH 78   P. N. Cheremisinoff, A. C. Morresi: *Air Pollution Sampling and Analysis Deskbook.* Ann Arbor Science Publ., Ann Arbor, MI, 1978.

CL 79   P. Cleij, A. Dijkstra: *Fresenius' Z. Anal. Chem.* **298**, 97 (1979).

CO 71   W. J. Conover: *Practical Nonparametric Statistics*, p. 302, Wiley, New York, 1971.

CO 77   W. B. Cochran: *Sampling Techniques*, Wiley, New York, 1977.

CO 81   D. C. Cornish, G. Jepson, M. J. Smurthwaite: *Sampling Systems for Process Analyzers.* Butterworth, London, 1981.

CO 85   I. A. Cowe, J. W. McNicol: *Appl. Spectrosc.* **39** (2), 257–266 (1985).

CU 68   L. A. Currie: *Anal. Chem.* **40**, 586 (1968).

DA 63   N. Dalkey, O. Helmer: *Manage. Sci.* **9**, 458 (1963).

DA 70   D. G. S. Davies: *Res. Dev. Manage.* **1**, 22 (1970).

DA 79   K. Danzer, K. Doerffel, H. Erhardt, M. Grissler, G. Ehrlich, P. Gadow: *Anal. Chim. Acta* **105**, 1 (1979).

DA 84   O. L. Davies, P. L. Goldsmith: *Statistical Methods in Research and Production.* Longman, London, 1984.

DE 73   S. N. Deming, S. L. Morgan: *Anal. Chem.* **45** (3), 278A (1973).

DE 78   S. N. Deming, L. R. Parker: *CRC Crit. Rev. Anal. Chem.* **7**, 187 (1978).

DI 69   W. J. Dixon, F. J. Massey: *Introduction to Statistical Analysis.* McGraw-Hill, New York, 1969.

DI 80   C. B. M. Didden, H. N. J. Poulisse: *Anal. Lett.* **13**, A11 (1980).

DI 84   W. R. Dillon, M. Goldstein: *Multivariate Analysis: Methods and Applications.* Wiley, New York, 1984.

DO 75   K. Doerffel: *Euroanalysis II.* Masson, Paris, 1975.

DY 74   A. Dyer: *Biometrika* **61**, 185 (1974).

EC 80   K. Eckschlager, V. Štěpánek: *Collect. Czech. Chem. Commun.* **45**, 2516–2523 (1980).

EC 85   K. Eckschlager, V. Štěpánek: *Analytical Measurement and Information.* Research Studies Press Ltd., Wiley, New York, 1985.

EN 73   A. W. England: *Sampling, Standards and Homogeneity,* ASTM STP 540, pp. 3–15. American Society of Testing Materials, Philadelphia, 1973.

FA 77   J. D. Fasset, J. R. Roth, G. H. Morrison: *Anal. Chem.* **49**, 2322 (1977).

FE 69   R. W. Fennel, T. S. West: *Pure Appl. Chem.* **18**, 439 (1969).

FI 70   A. F. Findeis, M. K. Wilson, W. W. Meinke: *Anal. Chem.* **42** (7), 26A (1970).

FO 71   J. W. Forrester: *World Dynamics.* Wright Allen Press, Cambridge, MA, 1971.

FO 85   D. L. Fox: *Anal. Chem.* **57**, 226R (1985).

FR 69   R. D. B. Fraser, E. Suzuki: *Anal. Chem.* **41**, 37 (1969).

FR 89   D. S. Frankel, R. J. Olson, S. L. Frankel, S. W. Chisholm: *Cytometry* **10**, 540–550 (1989).

GO 08   W. S. Gosset: *Biometrika* **6**, 1 (1908).

GO 72   G. Gottschalk: *Fresenius' Z. Anal. Chem.* **258**, 1 (1972).

GO 74   R. Goulden: *Analyst* **99**, 929 (1974).

GO 76   Z. Govindarajulu: *Comm. Stat. Theory Methods* A5, 429 (1976).

GO 88   D. Goulder, T. Blaffert, A. Blokland, L. Buydens, A. Chhabra, A. Cleland, N. Dunand, H. Hindriks, G. Kateman, H. van Leeuwen, D. Massart, M. Mulholland, G. Musch, P. Naish, A. Peeters, G. Postma, P. Schoenmakers, M. de Smet, B. Vandeginste, J. Vink: *Chromatographia* **26**, 237–243 (1988).

GR 63   P. M. E. M. van der Grinten: *Control Eng.* **10** (10), 87 (1963).

GR 65   P. M. E. M. van der Grinten: *J. Instrum. Soc. Am.* **12** (1), 48 (1965).

GR 66   P. M. E. M. van der Grinten: *J. Instrum. Soc. Am.* **13**(2), 58 (1966).

GR 85   J. J. de Gruijter, B. A. Marsman: In *Soil Spatial Variability* (D. R. Nelson, J. Bouma, Eds.), p. 150. Soil Survey Institute, Wageningen, The Netherlands, 1985.

GR 89   H. Grahn, N. M. Szeverenyi, N. W. Roggenbuck, F. Delagio, P. Geladi: *Chemom. Intell. Lab. Syst.* **5**, 311–322 (1989).

GY 82   M. Gy: *Sampling of Particulate Materials: Theory and Practice.* Elsevier, Amsterdam, 1982.

GY 86   P. M. Gy: *Anal. Chim. Acta* **190**, 13 (1986).

HA 69   P. D. Hall: In *Technological Forecasting and Corporate Strategy* (G. S. C. Willis, G. Ashton, B. Taylor, Eds.). Bradford University Press, Bradford, UK, 1969.

HA 74a   W. E. Harris, B. Kratochvil: *Anal. Chem.* **46**, 313 (1974).

HA 74b   A. den Harder, L. de Galan: *Anal. Chem.* **46**, 1464 (1974).

HA 80   S. H. Harrison, R. Zeilser: NBS Internal Report 80-2164, p. 66. National Bureau of Standards, Washington, DC, 1980.

HI 76   T. Hirschfeld: *Anal. Chem.* **48** (1), 16A (1976).

HO 76   W. J. Horwitz: *J. Assoc. Off. Anal. Chem.* **59**, 238 (1976).

HO 78   C.-N. Ho, G. D. Christian, E. R. Davidson: *Anal. Chem.* **50**, 1108–1113 (1978).

IL 85   F. K. Illingworth: *Trends Anal. Chem.* **4** (5), ix (1985).

IN 66   International Organization for Standardization Technical Committee 69: ISO/TC 69, 1966.

IN 73   C. O. Ingamells, P. Switzer: *Talanta* **20**, 547 (1973).

IN 74a   C. O. Ingamells: *Talanta* **21**, 141 (1974).

IN 74b   J. D. Ingle, Jr.: *J. Chem. Educ.* **51**, 100 (1974).

IN 76   C. O. Ingamells: *Talanta* **23**, 263 (1976).

IN 82   J. Inczédy: *Talanta* **29**, 643 (1982).

IN 87   International Organization for Standardization: *Quality Management and Quality Assurance Standards: Guidelines for Selection and Use*, 1987-03-15. International Organization for Standardization, 1987.

IV 86   A. Ivaska: *Anal. Chim. Acta* **190**, 89–97 (1986).

JA 83   T. A. H. M. Janse, G. Kateman: *Anal. Chim. Acta* **150**, 219–231 (1983).

JA 84   T. A. H. M. Janse, G. Kateman: *Anal. Chim. Acta* **159**, 181–198 (1984).

KA 61   R. E. Kalman, R. S. Bucy: *J. Basic. Eng.* **83D**, 95 (1961).

KA 67   H. Kahn, A. J. Wiener: *The Year 2000: A Framework for Speculation on the Next Thirty Three Years.* Macmillan, New York, 1967.

KA 70   H. Kaiser: *Anal. Chem.* **42** (2), 24A (1970).

KA 73   H. Kaiser: *Methodicum Chimicum*, Vol. I, Part 1, pp. 1–20. Thieme, Stuttgart, 1973.

KA 78   G. Kateman, P. J. W. M. Müskens: *Anal. Chim. Acta* **103**, 11–20 (1978).

KA79   G. Kateman, A. Dijkstra: *Fresenius' Z. Anal. Chem.* **247**, 249 (1979).

KA 81   G. Kateman, F. J. Pijpers: *Quality Control in Analytical Chemistry.* Wiley, New York, 1981.

KA 85   G. Kateman, P. F. A. van der Wiel, T. A. H. M. Janse, B. G. M. Vandeginste: *CLEOPATRA.* Elsevier Scientific Software, Amsterdam, 1985.

KA 90   G. Kateman: *Analyst* **115**, 487–493 (1990).

KE 88   L. H. Keith (Ed.): *Principles of Environmental Sampling.* American Chemical Society, Washington, DC, 1988.

KE 91   H. R. Keller, D. L. Massart, J. P. Brans: *Chemom. Intell. Lab. Syst.* **11**, 175–189 (1991).

KI 89   H. M. Kingston: *Anal. Chem.* **61**, 1381A (1989).

KL 75   L. A. Klimko, C. E. Antle: *Commun. Stat.* **4**, 1009 (1975).

KL 88   J. Klaessens, T. Saris, B. Vandeginste, G. Kateman: *J. Chemom.* **2**, 49–65 (1988).

KL 89   J. Klaessens, J. Sanders, B. Vandeginste, G. Kateman: *Anal. Chim. Acta* **222**, 19–34 (1989).

KO 71   D. Kortland, R. L. Zwart: *Chem. Eng.* No. 1, p. 66 (1971).

KO 75   B. R. Kowalski: *J. Chem. Inf. Comput. Sci.* **15**, 201 (1975).

KO 77   B. R. Kowalski (Ed.): *Chemometrics: Theory and Application*, ACS Symp. Ser. No. 52. American Chemical Society, Washington, DC, 1977.

KO 84   B. R. Kowalski (Ed.): *Chemometrics: Mathematics and Statistics in Chemistry.* Reidel, Dordrecht, The Netherlands, 1984.

KO 91   S. P. Koinis, A. T. Tsatsas, D. F. Katakis: *J. Chemom.* **5**, 21–35 (1991).

KR 81   B. Kratochvil, J. K. Taylor: *Anal. Chem.* **53**, 924A (1981).

KR 82   B. G. Kratochvil, J. K. Taylor: *NBS Tech. Note (U.S.)* **1153** (1982).

KR 84   B. Kratochvil, D. Wallace, J. K. Taylor: *Anal. Chem.* **56**, 114R (1984).

KU 79   H. H. Ku: *NBS Spec. Publ. (U.S.)* **519**, 1 (1979).

KU 85 D. A. Kutz: *ACS Symp. Ser.* **284** (1985).

KV 88 O. M. Kvalheim: *Chemom. Intell. Lab. Syst.* **4**, 11–25 (1988).

LE 50 S. Levey, E. R. Jennings: *Am. J. Clin. Pathol.* **20**, 1059 (1950).

LE 71 F. A. Leemans: *Anal. Chem.* **43** (11), 36A (1971).

LE 91a J. A. van Leeuwen, L. M. C. Buydens, B. G. M. Vandeginste, G. Kateman, P. J. Schoenmakers, M. Mulholland: *Chemom. Intell. Lab. Syst.* **10**, 337–347 (1991).

LE 91b J. A. van Leeuwen, L. M. C. Buydens, B. G. M. Vandeginste, G. Kateman, P. J. Schoenmakers, M. Mulholland: *Chemom. Intell. Lab. Syst.* **11**, 37–55 (1991).

LI 67 H. W. Lilliefors: *J. Am. Stat. Assoc.* **62**, 399 (1967).

LI 73 C. Liteanu, I. Rica: *Mikrochim. Acta*, p. 745 (1973).

LI 75 C. Liteanu, I. Rica: *Mikrochim. Acta* No. 2, p. 311 (1975).

LO 83 G. L. Long, J. D. Winefordner: *Anal. Chem.* **55**, 712A (1983).

LO 84 A. Lorber: *Anal. Chem.* **56**, 1004–1010 (1984).

LO 91 J. R. Long, H. T. Mayfield, M. V. Henley, P. R. Kromann: *Anal. Chem.* **63**, 1256–1261 (1991).

MA 51 F. J. Massey, Jr.: *J. Am. Stat. Assoc.* **46**, 69 (1951).

MA 73 D. L. Massart: *J. Chromatogr.* **79**, 157 (1973).

MA 75 D. L. Massart, L. Kaufman: *Anal. Chem.* **47**, 1244A (1975).

MA 77 D. L. Massart, H. de Clerq, R. Smits: *Euroanalysis II*: Masson, Paris, 1977.

MA 78a E. R. Malinowski: *Anal. Chim. Acta* **103**, 339–354 (1978).

MA 78b D. L. Massart, A. Dijkstra, L. Kaufman: *Evaluation and Optimization of Laboratory Methods and Analytical Procedures*. Elsevier, Amsterdam, 1978.

MA 88a E. R. Malinowski: *J. Chemom.* **3**, 49–60 (1988).

MA 88b D. L. Massart, B. G. M. Vandeginste, S. N. Deming, Y. Michotte, L. Kaufman: *Chemometrics: A Textbook*. Elsevier, Amsterdam, 1988.

MA 89 H. Martens, T. Naes: *Multivariate Calibration*. Wiley, New York, 1989.

MA 91 D. C. Mattes: *Chemom. Intell. Lab. Syst. Lab. Inf. Manage.* **13**, 3–13 (1991).

MB 83 A. B. McBratney, R. Webster: *J. Soil Sci.* **34**, 137–162 (1983).

MC 43 W. S. McCulloch, W. H. Pitts: *Bull. Math. Biophys.* **7**, 89–93 (1943).

MD 87 R. D. McDowall: *Laboratory Information Management Systems*. Sigma Press, Wilmslow, UK, 1987.

MF 70 E. F. McFarren, R. J. Lisha, J. H. Parker: *Anal. Chem.* **42**, 358–365 (1970).

MI 69 M. Minsky, S. Papert: *Perceptrons: An Introduction to Computational Geometry.* MIT Press, Cambridge, MA, 1969.

MI 77 D. Midgley: *Anal. Chem.* **49**, 510–512 (1977).

MI 86 G. Mitra (Ed.): *Computer Assisted Decision Making*, p. IX. Elsevier, Amsterdam, 1986.

MI 87 P. Minkkinen: *Anal. Chim. Acta* **196**, 237 (1987).

MI 88 J. C. Miller, J. N. Miller: *Statistics for Analytical Chemistry.* Ellis Horwood, Chichester, 1988.

MI 89 P. Minkkinen: *Chemom. Intell. Lab. Syst.* **7**, 189 (1989).

MO 76 D. F. Morrison: *Multivariate Statistical Methods.* McGraw-Hill, New York, 1976.

MU 78a P. J. W. M. Müskens, G. Kateman: *Anal. Chim. Acta* **103**, 1–9 (1978).

MU 78b P. J. W. M. Müskens: *Anal. Chim. Acta* **103**, 445–457 (1978).

MY 82 D. E. Myers: *Math. Geol.* **14**, 249–257 (1982).

NA 63 M. G. Natrella: *Experimental Statistics*, N.B.S. Handbook No. 91. U.S. Govt. Printing Office, Washington, DC, 1963.

NA 91 R. A. Nadkarni: *Anal. Chem.* **63**, 675A–682A (1991).

NE 81 J. D. Nelson, R. C. Ward: *Groundwater* **19**, 617 (1981).

OE 82 Organization of Economic Co-operation and Development: *Good Laboratory Practice in the Testing of Chemicals.* OECD, Paris, 1982.

OW 62 D. B. Owen: *Handbook of Statistical Tables*, pp. 413–425. Addison-Wesley, Reading, MA, 1962.

PE 66 D. C. Pelz, F. M. Andrews: *Scientists in Organizations.* Wiley, New York, 1966.

PI 85 J. W. Pijper: *Anal. Chim. Acta* **170**, 159 (1985).

PL 46 R. L. Plackett, J. P. Burman: *Biometrika* **33**, 305 (1946).

PO 79 H. N. J. Poulisse: *Anal. Chim. Acta* **112**, 361 (1979).

PO 92 G. J. Postma, G. Kateman: *Anal. Chim. Acta* **265**, 133 (1992).

RO 62 F. Rosenblatt: *Principles of Neurodynamics.* Spartan Books, Washington, DC, 1962.

RO 72 B. Roy: *Rev. METRA* **11**, 121–131 (1972).

RO 90 E. W. Robb, M. E. Munk: *Mikrochim. Acta* No. 1, pp. 131–155 (1990).

RU 89 S. C. Rutan: *Chemom. Intell. Lab. Syst.* **6**, 191–201 (1989).

SA 78	L. Sachs: *Angewandte Statistik*. Springer-Verlag, Berlin, 1978.
SC 66	A. C. Schram: *Tijdschr. Effic. Doc.* **36**(9), 511 (1966).
SC 73	E. I. Schuster: *J. Am. Stat. Assoc.* **68**, 713 (1973).
SC 77a	G. J. Scilla, G. H. Morrison: *Anal. Chem.* **49**, 1529 (1977).
SC 77b	B. Schmidt: *Fresenius' Z. Anal. Chem.* **287**, 157 (1977).
SE 79	P. F. Seelig, H. N. Blount: *Anal. Chem.* **51**, 327 (1979).
SH 31	W. A. Shewhart: *Economic Control of the Quality of Manufactured Product*. Van Nostrand, New York, 1931.
SH 49	C. E. Shannon, W. Weaver: *The Mathematical Theory of Communication*. University Press of Illinois, Urbana, 1949.
SH 65	S. S. Shapiro, M. B. Wilk: *Biometrika* **52**, 591 (1965).
SH 68	S. S. Shapiro, M. B. Wilk, J. H. Chen: *J. Am. Stat. Assoc.* **63**, 1343 (1968).
SH 72	S. S. Shapiro, R. S. Francis: *J. Am. Stat. Assoc.* **67**, 215 (1972).
SH 86	M. A. Sharaf, D. L. Illman, B. R. Kowalski: *Chemometrics*. Wiley, New York, 1986.
SI 56	S. Siegel: *Nonparametric Statistics for the Behavioral Sciences*, p. 44. McGraw-Hill, New York, 1956.
SK 92	B. Skagerberg, J. F. McGregor, C. Kisparissides: *Chemom. Intell. Lab. Syst.* **14**, 341–356 (1992).
SM 80	H. C. Smit: *Anal. Chim. Acta* **122**, 201 (1980).
SM 81	R. Smith, G. V. James: *The Sampling of Bulk Materials*. Royal Society of Chemistry, London, 1981.
SM 92	J. R. M. Smits, L. W. Breedveld, M. W. J. Derksen, G. Kateman, H. W. Balfoort, J. Snoek, J. W. Hofstraat: *Anal. Chim. Acta* **258**, 11–25 (1992).
SM 93	J. R. M. Smits, P. Schoenmakers, A. Stehmann, F. Sijstermans, G. Kateman: *Chemom. Intell. Lab. Syst.* **18**, 27–39 (1993).
SO 76	R. K. Som: *A Manual of Sampling Techniques*. Heinemann, London, 1976.
ST 70	M. A. Stephens: *J. R. Stat. Soc.* **B32**, 115 (1970).
ST 74	M. A. Stephens: *J. Am. Stat. Soc.* **69**, 730 (1974).
ST 87a	P. J. Stuyfzand: *Trends Anal. Chem.* **6** (2), 50 (1987).
ST 87b	R. Sturgeon, S. S. Berman: *CRC Crit. Rev. Anal. Chem.* **18** (3), 209 (1987).
ST 91	A Stein, L. C. A. Corsten: *Biometrics* **47**, 575–587 (1991).
SV 68	V. Svoboda, R. Gerbatsch: *Fresenius' Z. Anal. Chem.* **242**, 1 (1968).

TH 84 P. C. Thijssen, S. M. Wolfrum, G. Kateman, H. C. Smit: *Anal. Chim. Acta* **156**, 87–101 (1984).

TH 85 P. C. Thijssen, G. Kateman, H. C. Smit: *Anal. Chim. Acta* **173**, 265–272 (1985).

VA 74 I. R. Vaananen, S. Kivirikko, J. Koskenniemi, J. Koskimies, A. Relander: *Methods Inf. Med.* **13** (3), 158 (1974).

VA 75 B. G. M. Vandeginste, L. de Galan: *Anal. Chem.* **47**, 2124 (1975).

VA 77a B. G. M. Vandeginste: *Anal. Lett.* **10** (9), 661 (1977).

VA 77b B. G. M. Vandeginste, T. A. H. M. Janse: *Fresenius' Z. Anal. Chem.* **286**, 321 (1977).

VA 79 B. G. M. Vandeginste: *Anal. Chim. Acta* **112**, 253 (1979).

VA 80 B. G. M. Vandeginste: *Anal. Chim. Acta* **122**, 435 (1980).

VI 69 J. Visman: *Mater. Res. Stand.* **9** (11), 8 (1969).

VI 71 J. Visman, A. J. Duncan, M. Lerner: *Mater. Res. Stand.* **11** (8), 32 (1971).

VI 72 J. Visman: *J. Mater.* **7**, 345 (1972).

VO 81 J. G. Vollenbroek, B. G. M. Vandeginste: *Anal. Chim. Acta* **133**, 85 (1981).

WA 47 A. Wald: *Sequential Analysis.* Wiley, New York, 1947.

WE 74 J. O. Westgard, R. N. Carey, S. Wold: *Clin. Chem. (Winston-Salem, N.C.)* **47**, 825–833 (1974).

WE 80 R. Webster, T. M. Burgess: *J. Soil Sci.* **31**, 505–524 (1980).

WE 83 R. Webster, T. M. Burgess: *Agric. Water Manage.* **6**, 111 (1983).

WE 86 R. Wellmann, G. Wünsch: *Trends Anal. Chem.* **5**(6), 138–141 (1986).

WE 89 R. Webster, M. A. Oliver: *J. Soil Sci.* **40**, 497–512 (1989).

WE 93 R. Wehrens, P. van Hoof, L. Buydens, G. Kateman, M. Vossen, W. H. Mulder, T. Bakker: *Anal. Chim. Acta* **271**, 11–24 (1993).

WI 88 S. Williams (Ed.) *Official Methods of the Association of Official Analytical Chemists,* 15th ed. AOAC, Arlington, VA, 1988.

WO 79 T. C. Woodis, J. H. Holmes, B. Hunter, F. J. Johnson: *J. Assoc. Off. Anal. Chem.* **62**, 733 (1979).

WO 88 R. Wolters, G. Kateman, *Anal. Chim. Acta* **207**, 111–123 (1988).

YA 86a S. R. Yates, A. W. Warrick, D. E. Myers: *Water Resour. Res.* **22**, 615–621 (1986).

YA 86b S. R. Yates, A. W. Warrick, D. E. Myers: *Water Resour. Res.* **22**, 623–630 (1986).

YO 72   W. J. Youden: *Statistical Techniques for Collaborative Tests.* Association of Official Analytical Chemists, Washington, DC, 1972.

YO 75   W. J. Youden, E. H. Steiner: *Statistical Manual of the Association of Official Analytical Chemists.* AOAC, Washington, DC, 1975.

ZU 91   J. Zupan, J. Gasteiger: *Anal. Chim. Acta* **248**, 1–30 (1991).

INDEX

307